Sol-Gel Synthesis and Processing

Related titles published by the American Ceramic Society:

Sol-Gel Processing of Advanced Materials
Edited by Lisa C. Klein, Edward J.A. Pope, Sumio Sakka, and James L. Woolfrey
© 1998, ISBN 1-57498-042-4

For information on ordering titles published by the American Ceramic Society, or to request a publications catalog, please contact our Customer Service Department at 614-794-5890 (phone), 614-794-5892 (fax), <customersrvc@acers.org> (e-mail), or write to Customer Service Department, 735 Ceramic Place, Westerville, OH 43081. Visit our on-line book catalog and web site at <www.acers.org>.

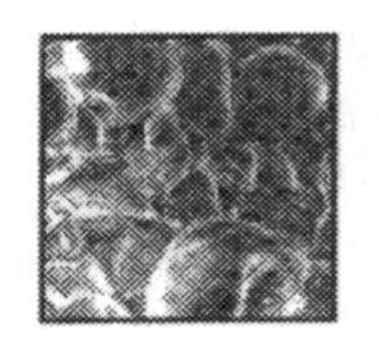

Ceramic Transactions
Volume 95

Sol-Gel Synthesis and Processing

Edited by

Sridhar Komarneni
The Pennsylvania State University

Sumio Sakka
Fukui University of Technology

Pradeep P. Phulé
University of Pittsburgh

Richard M. Laine
University of Michigan

Published by

The American Ceramic Society
735 Ceramic Place
Westerville, Ohio 43081

Proceedings of the International Symposium on Sol-Gel Processing, held at the 100th Annual Meeting of The American Ceramic Society in Cincinnati, Ohio, May 3–6, 1998.

Copyright 1998, The American Ceramic Society. All rights reserved.

No part of this book may be reproduced, stored in a retrieval system, or transmitted in any form or by any means, electronic, mechanical, photocopying, microfilming, recording, or otherwise, without written permission from the publisher.

Permission to photocopy for personal or internal use beyond the limits of Sections 107 and 108 of the U.S. Copyright Law is granted by the American Ceramic Society, provided that the base fee of US$5.00 per copy, plus US$.50 per page, is paid directly to the Copyright Clearance Center, 222 Rosewood Dr., Danvers, MA 01923, USA. The fee code for users of the Transactional Reporting Service for *Ceramic Transactions Volume 95* is 1-57498-063-7/98 $5.00+$.50. This consent does not extend to other kinds of copying, such as copying for general distribution, for advertising or promotional purposes, or for creating new collective works. Requests for special photocopying permission and reprint requests should be directed to the Director of Publications, The American Ceramic Society, 735 Ceramic Place, Westerville OH 43081, USA.

COVER PHOTO: "Morphology of alumina foams treated at 1300°C," is courtesy of G.S. Grader, Y. de Hazan, and G.E. Shter, and appears as figure 6 in their paper, "Ultra Light Ceramic Foams," which begins on page 161.

Library of Congress Cataloging-in-Publication Data
A CIP record for this book is available from the Library of Congress.

For information on ordering titles published by The American Ceramic Society, or to request a publications catalog, please call 614-794-5890.

1 2 3 4–01 00 99 98

ISSN 1042-1122
ISBN 1-57498-063-7

Contents

Porous Materials through Sol-Gel Processing

Basic and Applied Sol-Gel Science and Processing of Ceramics and Composites

Preface

This book is dedicated to Professor Rustum Roy of The Pennsylvania State University, who has been using the sol-gel process for making ceramics via mixing in solution, gelation, desiccation, and firing since 1948. Dr. Roy's review of the sol-gel process published in the *Journal of the American Ceramic Society* in 1956 became a citation classic.

Today, in many conferences on ceramics, a significant number of papers utilize the sol-gel process. This symposium was held as a special golden anniversary of Roy's use of sol-gel process at the centennial meeting of The American Ceramic Society. A special feature of the symposium was the opening historical session followed by the recent advances in sol-gel synthesis and processing. The symposium has covered topics on historical aspects of sol-gel; electroceramic films by sol-gel process; basic and applied sol-gel science; porous materials through sol-gel processing; and sol-gel processing of ceramics, glasses and composites. This symposium has brought together scientists from different disciplines such as ceramic science, materials science, chemistry, and chemical engineering, and it covered the synthesis and processing of many kinds of materials. A total of 45 papers has been presented from a large number of universities, government laboratories, and private industry in 12 countries. The 23 papers published in the proceedings have been reviewed and divided into four sections. The first section deals with historical aspects of sol-gel, the second section deals with the electroceramic films by sol-gel process, the third section deals with porous materials through sol-gel processing, and the fourth and final section deals with the basic and applied sol-gel science and processing of ceramics and composites.

Sridhar Komarneni
Sumio Sakka
Pradeep P. Phulé
Richard M. Laine

Historical Aspects of Sol-Gel Processing

THE SOL-GEL PROCESS IN CERAMIC SCIENCE: EARLY HISTORY OF DISCOVERY AND SUBSEQUENT DEVELOPMENT

Rustum Roy
The Pennsylvania State University
102 Materials Research Laboratory
University Park, PA 16802

ABSTRACT

This paper records the earliest history of the development of the sol-gel process as a means of making pure and extremely fine grained glass and ceramic materials starting from gels, originating from solutions (inorganic and organic). Such "solution sol-gel" science had as its original goal the achievement of precursor materials with maximum purity, compositional *homogeneity* and *reactivity* for research in ceramics. In technology, the major potential (which is not very large) has probably already been realized as far as bulk products are concerned.

Since 1982, the author has created a new goal for the sol-gel process: that of achieving materials of maximum *heterogeneity* in precursors; to which the term 'nanocomposities' was first applied by him. The latter part of this paper summarizes some of the fundamental reasoning for the use of monophasic and multiphasic gels and illustrates the emergence of new potential applications with several examples from real technology and recent work in the author's laboratory.

INTRODUCTION

This paper is not a review of the field. I was invited to present my personal involvement in the earliest history of the field and my perspective on it today. Every review on history presents in any case the perspective of the writer. I have written, earlier, such a review in Science[1]. There are many well known books, e.g. by Iler[2] which combine historical elements, and others which have much scientific background e.g. Scherrer and Brinker.[3] Each has its own goals and perspective. A recent review by Wood and Dislich[4] reviews SSG-technology. It is a fine review by very experienced, distinguished writers prepared for patent

To the extent authorized under the laws of the United States of America, all copyright interests in this publication are the property of The American Ceramic Society. Any duplication, reproduction, or republication of this publication or any part thereof, without the express written consent of The American Ceramic Society or fee paid to the Copyright Clearance Center, is prohibited.

purposes. The review goes back to ancient technologies vaguely manifest in archeological artifacts from 2000-4000 B.C. That review, because of the authors' perspective, omits the entire field of food, where humans for millennia daily encountered sols and gels from tofu and yogurt to milk itself. The scope of this present history is narrower. It is concerned with the making of ultrafine, reactive, very pure ceramic precursors of any composition, for research in ceramic science. That was my goal in 1947-48, when I became involved in this area. Our goals from the earliest years covered making such pre-cursors not only for the simplest unary systems: Al_2O_3 and SiO_2, but for forsterite, cordierite, feldspars, micas, clays with four and five components.

HISTORY OF SOLUTION-SOL-GEL SCIENCE

In the early nineteen-fifties in a comment at the presentation of a paper of mine my colleague Professor W.D. Kingery of M.I.T. jokingly remarked that I was attempting to create a new field of "solution ceramics." This was an accurate categorization of the topic I will discuss herein and the fact that it was regarded as far offbeat is definitively shown in the neglect of the topic by the materials research community *for two decades.* Materials research has advanced greatly by the creation of new materials and processes (and new instruments). The solution-sol-gel (SSG) process is probably the most widely used new process innovation in general ceramic science research, although its application in industry is minuscule compared to, say, glass ceramics. SSG-derived ceramics are a subset of novel materials made by atomic scale mixing in solution of the constituents. The history of the topic therefore has three parts: first the concept of mixing in solution, and second the utilization of the gelation as a step in the process to retain homogeneity, and third the shaping of the gel prior to firing into the final product. (see Ref. 1 for details)

Mixing in Solution

Immediately after WWII the glasses and ceramics, the properties of which were the most accurately determined, were the laboratory samples used in phase equilibrium studies of the most significant ceramic systems. The arcane technique for making such samples had been developed and handed down through a succession of the most distinguished workers at the Geophysical Laboratory of the Carnegie Institution in Washington, DC. The names of N. L. Bowen, J. W. Greig, J. F. Schairer, G. W. Morey, E. F. Osborn, are familiar to all as the principal contributors to understanding the high temperature thermodynamics of the main ceramic oxide systems.[5] I was inducted into this tradition of preparing homogeneous glasses and ceramic powders therefrom as the first Ph.D. student of E.F. Osborn who had moved to Penn State in 1946 and set up the nation's second major center for phase equilibrium research. The Geophysical Laboratory process for making maximally homogeneous glasses and ceramics therefrom consisted of the melting of carefully weighed, well-mixed mixtures of oxides and carbonates of the composition desired, quenching to a glass, crushing to a fine powder in steel mortars, removing iron carefully with a magnet, and re-melting. It was necessary to repeat this entire process three to five times to achieve acceptable homogenization. I completed the determination of the equilibrium diagram of the system Li_2O_3-Al_2O_3-SiO_2 using this method to make homogeneous glasses and fine powders. Not only was it extremely tedious, but as I started to apply it to my own new low temperature phase equilibrium study — of the system Al_2O_3-SiO_2-H_2O [(6)] — two other major limitations of the melting to glass process came into focus:

First: The compositional range over which homogeneous glassy, thence crystalline, phase(s) could be obtained was quite narrow (e.g. in Al_2O_3-SiO_2 only compositions from 2 up to 15% Al_2O_3 could be retained as glasses).

Second: The structural bias introduced into these precursor glassy phases by the use of the high melting temperatures tended to favor a certain set of metastable products (in the case of Al_2O_3-SiO_2, those with Al in IV c.n.).

Moreover, using mixtures of the end members i.e. quartz and corundum, led to literally zero reaction in the p-t range involved. It thus became obvious that an alternative route to the preparation of chemically homogeneous glasses and other ceramic powders was desirable, and I therefore set about examining those methods which achieved homogeneity on an atomic scale by putting all the ions into solution. By chance, in a three- or four-year period starting July 1, 1948, out of this effort, we had developed three areas each of which has had a lasting impact on ceramic science: the sol-gel technique, new hydrothermal reaction techniques, and applications of phase separation in titania-rich glasses to glass-ceramics.

Relation to Colloid Science

The science of colloids or sols focused the attention of the chemistry community on this peculiar state of matter: a macroscopically homogeneous, "permanent"* suspension of solid particles in a liquid. Yet, in 1948, the bridges between colloid chemistry and ceramics were essentially zero outside the field of clay forming. None of the great colloid chemists--Zsigmondy, Svedberg, Freundlich[7-9] or their U.S. counterparts Weiser, Milligan, Hauser[10] got involved in anhydrous ceramic materials, although Weiser and Milligan worked extensively with oxide gels. Gels made the first link to ceramic technology via the patented methods for making SiO_2 and Al_2O_3 and TiO2 gels, including aerogels for various applications.[11 -12]

The next point of contact in applying colloid science to ceramics was in approaches to synthesizing clays.[13,14] Ewell and Insley made elaborate studies of co-precipitated (from sodium silicates and aluminum nitrates) gels of Al_2O_3 and SiO_2, purified by electrodialysis to remove Na as starting materials for kaolin synthesis. Raychaudhari and Dattu[15] published similar work on aluminosilicate and iron silicate precipitates from sols. A benchmark in the summary of research on sols and gels of silica and the silicates was R.K. Iler's "Baker Lectures" at Cornell.[2] Yet ceramics and glasses are hardly mentioned in the book, with the exception of the chemistry of synthesizing micas and clays (hydroxylated ceramic phases) via the SSG, route as noted above.

The first application of sol-gel methods for the production of chemically homogeneous anhydrous ceramics-bulk glasses in his case-appeared in a single paper by de Korosy[16] who also used the sodium aluminate-sodium silicate co-precipitation as the basis for making a gel precursor for glass.

Based on this background, our original efforts in 1948 were intended to make pure, anhydrous, homogeneous, crystalline and glassy one and <u>multi-component</u> ceramic phases. For the Al_2O_3-SiO_2 system it appeared possible, since one ion could be present as a cation and the other as an anion as Ewell and Insley[14] had already shown. That same process was also starting to be widely used in the <u>technology</u> of noncrystalline though anhydrous Al_2O_3-SiO_2 gel beads used as cracking catalysts. This was the method I first followed in trying to make homogeneous glasses in the system Li_2O-Al_2O_3-SiO_2, MgO-Al_2O_3-SiO_2, and found that with Na-salts as starting materials significant amounts of alkali and changes in Al/Si ratio were difficult to avoid. Even though I performed elaborate electrodialysis of the gels, purity and stoichiometry were problematic. Moreover, it soon became evident that neither could one obtain soluble salts of most high-

*My friend, Professor Sir John Thomas, then President of the Royal Institution, showed me recently the vial containing a purple fluid, which was the gold sol prepared by Faraday himself well over 100 years before, and is still stable.

temperature oxides, nor was it possible to 'co-precipitate' all cations under the same Eh, pH conditions (not to mention the segregation 'via crystallization' problem during desiccation). It should be noted that in 1948 not a single stable oxide sol was available commercially. Ludox became available in the mid-fifties and we were one of DuPont's eager customers. Hence, the key innovation in establishing the solution-sol-gel process for making pure ceramic phases both glassy and crystalline was the author's introduction in 1949 of the more universal process *of using organometallics as sources of the major cations-e.g.* using ethyl orthosilicate and Al-isopropoxide in the Al_2O_3-SiO_2-H_2O study.(6) The method was successfully generalized to a three-component oxide system, MgO-Al_2O_3-SiO_2, by D.M. Roy and R. Roy.(17) Over the next decade this method for making precisely compositionally controlled glasses and ceramics was applied to virtually any composition even in 5-, 6- or 7-component systems by the author and his colleagues and students. The generalization proved so successful that dozens of graduate students started to use the "gel method" universally in the large number of mineral synthesis and phase equilibrium studies carried out in Penn State's geochemistry and ceramic technology departments in the fifties and sixties. A typical example of making the important ceramic phase, cordierite, ran as follows. The aliquot of prestandardized solution of TEOS in absolute alcohol was added to a solution made from the weighed amounts of the nitrates of Mg and Al dissolved in alcohol and water. This solution by adjusting pH and temperature (usually from 60°-90°C) gradually formed a sol and then, by appropriate temperature and time adjustment, a gel. This gel was desiccated to a xerogel, first on a water bath then typically at 500°-700°C. This yielded a noncrystalline solid (NCS) ultrafine, chemically pure powder of the cordierite or any other chosen composition. Such powders in our work were typically and routinely (a) melted to glass, (b) hydrothermally crystallized to fine hydrous or anhydrous, low temperature ceramic powders with a high degree of structural order unattainable by any other process, (c) crystallized dry to a ceramic aggregate (though *not* as a *shaped desired product).*

Starting in 1951, our work on systematic phase equilibria began to include the titanates and niobates, and the sol-gel method was further generalized in a wide variety of papers. As a source for Ti we used both $TiCl_4$ and titanium-butoxide in many cases, e.g. in the systems TiO_2-SiO_2,(18) BaO-TiO_2-SiO_2,(19) BaO-CaO-TiO_2-SiO_2,(20) etc. Many of these early oxide powders or xero-gels were used in "hydrothermal" work as reactive starting materials and the author's first brief review on the sol-gel process carried this connection in the title.(21) The goal which was a process for making nanoscale (in today's terminology) ultrapure, stoichiometric ceramic powders of virtually any composition had been reached. The title of my first review "Methods of making mixtures for both dry and wet phase equilibrium studies," accurately points to the goal we had achieved: Pure reactive ceramic precursors for scientific research, and that is where the major impact has been. This paper on the sol-gel process became the first Citation Classic in the field of ceramics and the first citation classic ever published (Fig. 1) in the *Journal of the American Ceramic Society,* showing objectively its impact on the field by 1987.

August 17, 1987 Volume 18 Number 33

CURRENT CONTENTS®

Engineering, Technology & Applied Sciences

INCLUDING
Acoustics • Aeronautics & Aerospace Engineering
Automation • Chemical Engineering
Civil Engineering • Computer Science & Technology • DP Applications • Electrical Engineering
Electronics • Energy
Environmental Engineering
Geotechnology • Industrial Engineering • Instrumentation
Laser Science & Technology
Marine Engineering • Materials Science • Mechanical Engineering
Mechanics • Metallurgy • Mining
Nuclear Engineering • Operations Research • Optics • Petroleum Engineering • Polymer Science
Robotics • Telecommunications
Textiles • Transportation Technology

Citation Classics®:

Roy R. Aids in hydrothermal experimentation: II, methods of making mixtures for both "dry" and "wet" phase equilibrium studies. *J. Amer. Ceram. Soc.* *39*:145-6, 1956.

Current Comments®:

Research and Dedicated Mentors Nourish Science Careers at Undergraduate Institutions

ISI®
Institute for Scientific Information®
3501 Market Street, Philadelphia, Pennsylvania 19104 U.S.A.

Figure 1: Cover of Current Contents issue, which records the fact that the original sol-gel article in the Journal of the American Ceramic Society, had been cited so frequently that it had attained the status of "Citation Classic." It was the first ever in that Journal and the first in the field of ceramics.

One disadvantage of the use of ethyl orthosilicate as the source of silicon was the possibility of loss of some of the vapor during the 75°-90°C hydrolysis, and gelation. It was originally for that reason that we experimented with the use of an alternate source of SiO_2 in the form of an inorganic SiO_2-sol ("Ludox" 35% of which consisted of 200Å particles of "SiO_2" dispersed in H_20 and stabilized by NH_4^+ had just come on the market) as a source of SiO_2. These and similar commercial sols of "Al_2O_3" and "ZrO_2" were soon often used in our gels. Indeed by the late 1950's and 1960's the use of inorganic sol precursors became the more common procedure in much of our laboratory-scale work, because they were much cheaper and more manageable and to the best of our findings gave products indistinguishable from those made from organometallics.

Although the materials research community totally neglected it, the 'gel' method for making fine powders of oxide mixtures was utilized widely by geoscientists in experimental petrology.[22-25]. By then we had established the following significant conclusions:

1. The terms in common use are defined and schematically illustrated below in Fig. 2.

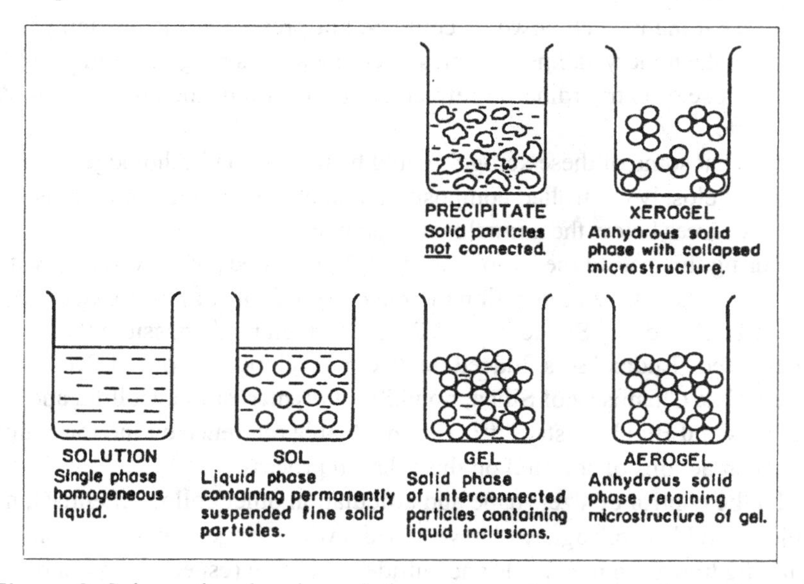

Figure 2: Schematics showing relations among the commonly used terms in this paper.

2. The solution-sol-gel process, using either organic precursors or very fine (2-20nm) sols of SiO_2, "Al_2O_3," ZrO2, etc., could be used to make ultrahomogeneous slightly hydrated or anhydrous, ceramic powders in virtually any single or multi-component ceramic system. We estimate that some 5000 such (nanosize ceramic powder) samples in dozens of complex ceramic systems were made- most of them still exist. For the historical record, Figs 3 and 4 show photographs of the actual samples of some of the earliest such powders nearly 50 years old. A color slide showing several hundreds of these in our present files was presented in the oral presentation.
3. That such powders, when heated dry in the range 500°-$1000^{\circ}C$ would typically yield anhydrous noncrystalline oxide materials [which rigorously cannot be called glasses] extending the compositional range of noncrystallinity far beyond that attainable by the more typical, liquid→solid (glassmaking), or vapor→solid routes. Indeed this range far exceeds what is possible by the new techniques of rapid quenching of melts at $\approx 10^{6\circ}$/sec.
4. That many such powders could be hot-pressed at modest temperatures to make noncrystalline ceramics of compositions ranging from SiO_2[26] to those of extraordinarily refractory compositions including ThO_2, ZrO_2, etc.
5. That some of these xerogels could be melted to ultrahomogeneous glasses of closely controlled composition and at temperatures hundreds of degrees below the normal melting range.

During this period the colloid chemists' interest began to converge with those of the ceramists in the preparation of *concentrated* sols of many oxides. It was the work of Bechtold and Snyder[27] and Rule[28] which made possible the concentrated "Ludox" sols. Indeed the later encyclopedic work of R.K. Iler entitled "The Chemistry of Silica, Solubility Polymerization, Colloid and Surface Properties and Biochemistry,"[28] is an invaluable reference on the formation, stability, reactions of sols and on the gelation process.

To the control of size, shape and articulation other colloid chemists, notably Matijevic and his colleagues[29] have added a wider range of compositions (including transition metal sols and sulfides) and size (especially monodisperse sols) and morphology.

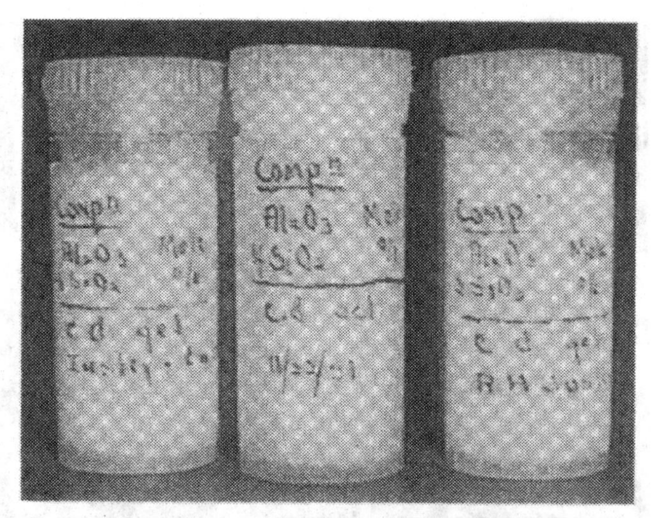

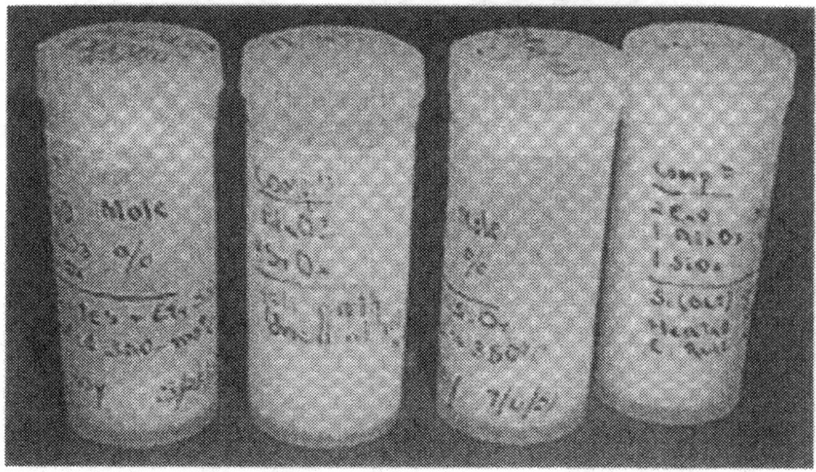

Figure 3: Photographs of bottles containing some of the actual earliest (1948-50) samples of xerogels in the systems Al_2O_3-SiO_2.

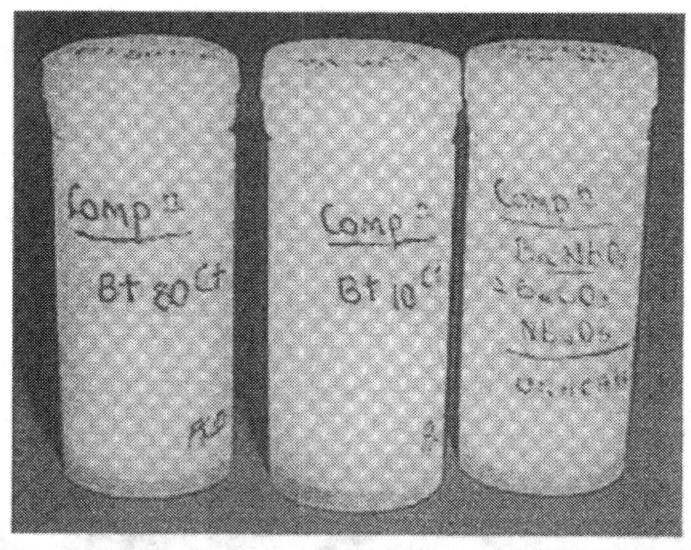

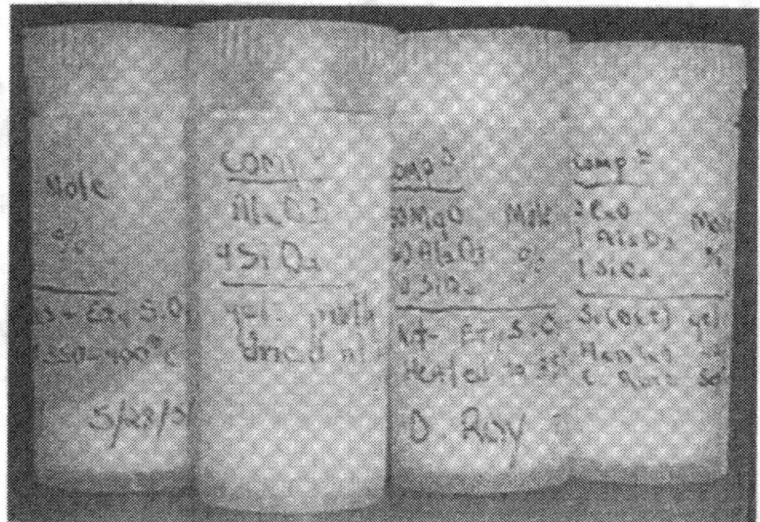

Figure 4: Photographs of earliest xerogels samples made multicomponents and involving electroceramics including titanates, niobates, etc. dating to the early fifties. A couple of thousand of such mixtures exist.

PROCESSES FOR MAKING CERAMICS VIA SSG

The principal advantage which has driven all solution-based approaches is the fact that by mixing all the components in solution atomic-molecular *homogeneity* is achieved immediately, instead of by the slow, energy-consuming process of thermally activated solid state diffusion (which for most common oxide ceramics means heating at 1200°-2000° for hours or days). Use of sols instantly increases the homogeneity down to 10-20 nm. Hence it is very appropriate to call the generic process the solution-sol-gel process (SSG).

Figure 4 shows the essentials of the generic SSG process and how it is used in the different technologies for making a variety of ceramic materials. The figure and legend are largely self- explanatory--only a few comments are in order:

Step 1. The "solution" which in a multi-component composition is a true ionic or molecular solution makes possible the atomic scale mixing which is the basis for the "ultra-homogenization" made possible by the SSG technique. The solvent for the vast majority of oxide gels is water and/or a short chain alcohol and the solutes may be either inorganic nitrates, chlorides, etc., or a wide variety of metal organic molecules.

Step 2. Conversion to a sol is accomplished by adjusting the activity of some species, H^+ and OH^- and other salts (or mutual precipitation in some cases such as in some aluminosilicates made from alkali silicates and aluminates), which results in the formation of a dispersed solid phase. In many cases a stable sol pre-formed via a Step I elsewhere is one of the starting materials for the next steps, so that there is no true solution step I at all. Yet in some such cases, we have shown by MASNMR that Al^{3+} ions, for example, from the solution can enter the tetrahedrally linked solid SiO2 particles.[30]

Step 3. Gelation of the sol is controlled principally by pH, concentrations, temperature and time. Manipulation of these parameters is an empirical procedure which has to be worked out for each composition and while the general conditions of pH and T are known for simple oxides such as SiO_2 and AIOOH the fine tuning of this step for different starting materials, especially new organic precursors, and multi-component systems has become a major area of current research (vide infra).

Step 4. This is *the* key step in the *technological* utilization of the SSG process-the *shaping of the gel to near final shape* (if not dimensions). In our work in the fifties and sixties we had chosen only one option-not to shape the gel deliberately but merely to desiccate it to powders or melt to glasses. Each separate family of SSG technological invention since, has basically consisted of a new shaping of the gel while still in the plastic condition into different shapes or forms, each one near to the respective final desired shape or form.

The diagram in Figure 5 shows the following important options for the final shape of the gel: spheres, small dice, fibers, thin sheets (= coatings) on a substrate. Each of these shapes, as we describe below, has led to a major new technology.

Step 5. Desiccation and heat treatment to a xerogel and finally to a glass or ceramic. The processing here obviously varies depending on the product being made, but in all cases the highly reactive nature of the xerogel assures that the time and temperature of reaction are both dramatically lowered when compared to conventional firing of ceramic bodies.

Up to the present, the maximum dimensions of any commercial densified ceramic product (not glass) made via the SSG process has been limited to a 1-2 mm in 3 dimensions (spheres or dice); a few microns in 2 dimensions and large in the other (in the case of fibers); and a few microns or less in one dimension and large many meters in the other two dimensions, as in the case of coatings.

The limitation in size of ceramic objects made by the SSG route is, of course, due to the necessity for extracting the liquid phase and maintaining coherence and integrity during that step. In spite of several hundreds of aggregate person-years of research by many industrial groups the problem of making large anhydrous objects, except by melting, has not been solved.

SOME TECHNOLOGIES BASED ON THE SSG ROUTE

The post-1970 interest in the *science* of the sol-gel process has often neglected, and still misses, the early scientific literature detailed above. But it also ignores the very large body of patent literature and actual technological practice in the SSG field. Many persons new to the field often appear to be totally unaware not only of the technology of the ancients and the earlier patents so ably summarized by Wood and Dislich[(4)], but also of SSG-derived products which have been on the market for years. We describe very briefly below the major classes of existing SSG *technologies.*

Fuel Pellets

In 1960, Mumpton and Roy[(31)] showed how to use the SSG process to make their highly reactive ceramic *powders* for the study of reactions and phase equilibria in the system UO_2-ZrO_2-ThO_2, In the mid and late sixties, the Oak Ridge group conceived and executed the first purposive shaping of the gel of these same compositions into small spheres[(32)] for use as nuclear fuel. The ingenious idea was merely to cause gelation to occur by change of pH while a drop of the solution of U or Th was falling slowly in a long column of a heated second organic liquid with which it was immiscible, but which caused

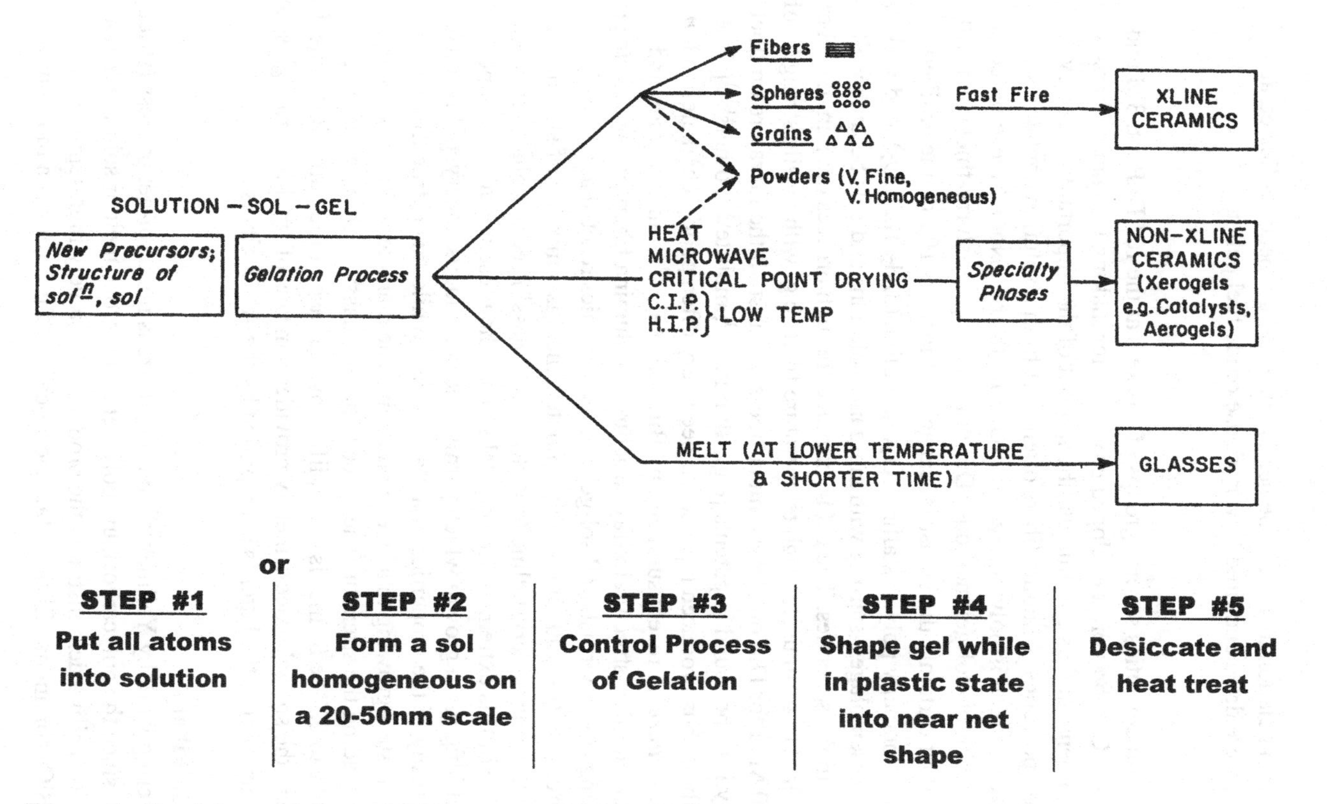

Figure 5: Breakdown of the total SSG process with its component steps.

hydrolysis and gelation to set in. The nearly spherical gel beads collected at the bottom, were dried and fired to very dense oxide fuel pellets

Fibers

In the mid-seventies several groups of workers at the ICI DuPont, 3M and Carborundum Corporations developed processes for making inorganic fibers-both as short filaments and as continuous fibers -- out of the important refractory ceramic compositions. Alumina fibers often start from a solution containing "$Al_2(OH)_5Cl$" which is polymerized by changing the pH. Not all these processes are strictly solution-sol-gel methods (DuPont uses slip- or slurry-thickened with an organic) but certainly the industrial capacity to make 5-10μ diameter fibers containing principally polycrystalline Al_2O_3, mullite ($3Al_2O_3.2SiO_2$), and various polyphasic assemblages, starting with inorganic solutions and sols became well established in the seventies. The gel is extruded through orifices or spun, the fibers dried and fired to form useful high temperature fibers e.g. with tensile strengths of 1300-2000 MPa for DuPont FP Alumina fibers. Almost all the literature on fiber technology is to be found in patents [e.g., alumina[33]; mullite[34]; others[35]]. The ceramic fibers thus produced have so far been very often poorly characterized *in the open literature.* A recent summary by Birchall et al.[36] is an exception and shows detailed microstructural x-ray structures of alumina fibers at various stages of processing. As in virtually all sol-gel derived ceramics reacted at relatively low temperatures (1000°-1200°C) for a few minutes, there is a mixture of submicron crystals with some noncrystalline (not strictly a "glass") second phase. It is the maintenance of this extremely fine crystal size by the presence of the second phase inhibiting grain growth which permits the continuous noncrystalline phase to exhibit many of the properties of a glassy fiber. While this is true there is no chance of crystals growing to a size approaching the diameter of the fiber and thereby degrading the strength. The percentage of "glassy" phase varies from nearly 100% to 0. The former is exemplified by the case of the pure SiO_2 optical fibers where the sol-gel process merely provides a means of "melting" to a glass at a lower temperature; the latter is exemplified by the Al_2O_3 fibers.

Non-Oxide Fibers

The pioneering work by Yajima[37,38] on SiC fibers is not a sol-gel process (strictly speaking), since the polycarbosilane polymer is a viscous melt or solution, not a sol. In 1976, Yajima demonstrated the production of a very high-temperature ceramic (SiC) continuous fiber. In fact the so-called "SiC" fiber is a multiphasic fiber consisting of very fine, poorly formed SiC crystals in a noncrystalline matrix of Si-C-O-H-N with the SiC content often no more than 90-95 atomic percent.

Coatings

The Schott Company in Jena had been working on the SG coating of glass since before WWII [39] although again most of the literature is in patents. A series of recent papers by Dislich et al.[40,41] provide some details on the production and detailed studies of thin (≅1μ) coatings of SiO_2, TiO_2, etc., on glass for reflective and anti-reflective effects. In recent years some 10^5 m^2 of window glass have been coated annually by this one company alone by the SSG process (most of the gold-colored windows in cities are coated by this process). Thus while useful thin ceramic coatings are common commercial articles, the many efforts which have been made all over the world to make useful *thicker* (>10-25g) sol-gel derived coatings using both organic and inorganic starting materials have failed. Yoldas, in his work, was successful in making high-performance coatings of "TiO_2" up to approximately lμ thickness, by carefully controlling the polymerization of the inorganic gel.[42]

Bulk Ceramic / Glass and Abrasive Grain

It has been the goal of hundreds of researchers in dozens of companies and some universities to make dense bulk ceramics from a gel. This has eluded all efforts so far. Making bulk glasses on the other hand is a trivial experiment of merely melting large batches of xerogels. Two or three significant efforts may be noted. In the early seventies the Owens-Illinois Company explored experimentally the costs and benefits (mainly ultrahomogeneity and lower melting temperatures) of making full-scale tank melts via the SSG process.[Thomas[43]] Just before the energy price increase of the oil shock, the process was judged to be uneconomical. Bausch and Lomb Optical Co. offered in the sixties to develop a custom-melting of the ultrahomogeneous optical glasses (such as laser-hosts) based on SSG techniques where applicable--it proved not to be viable. More recently Harrison and Pope[44] have taken effective advantage of the lower melting temperatures (1300°) to obtain a more chemically resistant aluminosilicate radioactive waste glass than the reference borosilicate glass. By traditional melting methods it is simply impossible to retain many of the volatile elements in radioactive waste at the temperatures required to melt and homogenize the <u>aluminosilicate</u> glass made from traditional starting materials. Yet it has never been adopted for trial.

The maximum dimension ceramic technological product which is SSG-derived, is the new "Al_2O_3," abrasive grain for grinding wheels developed by Leitheiser and

Sowman.[45] The advantages of this process over the traditional process for making the same product, illustrated in the following schema, show why the scientific community built up so many hopes for the SG process based on this success.

Traditional Process	**Mix. Bayer → process "alumina" with ~ 1% CaO**	**Melt in electric arc furnace at 2500°C for 30 min.**	**Solidify into 1-meter size ingots**	**Grind ingots to mm-size grain**	
SSG Route	**Mix 93% Al and 7 % Mg as a nitrate solution**	**Adjust pH → sol →**	**Pour into 1-2 cm. deep layer → Gel →**	**Dice dried gel in tray; fire at 1350°C for 1-5 min.**	**Separate mm-size grain**

Very recently in our own laboratories, we have taken this one step further by using microwave sintering to make such abrasive grain with superior properties in 15 minutes.[46]

A PERSPECTIVE ON CURRENT SOL-GEL RESEARCH

The recent almost "fashionable" concentration of research on ceramics derived from sol-gel may be characterized in several ways. First, besides the discovery and development of the science in the author's laboratory, the early work was conducted largely in industry and reported in patents; now it is largely in universities and government laboratories and appears as papers. The early work was focused on the *products* or new *material,* the newer work is focused on the science behind the already well-established processes and hence is not directly related to this review. The recent literature is enormous. Indeed in some national ceramics meetings sometimes over a quarter of all the papers use the SSG process in some way. No attempt can be or is made here to summarize this work systematically. The interested reader is referred to the proceedings volumes of the several successive international symposia which occur every few months.[47] Instead, in the following brief overview, I will try to connect these trends only in the first decade of this more recent work,[48]* to the process steps of the SSG process shown in Fig. 5.

---

*I may appear to ignore some recent advances in each of these steps, but the purpose of *this* review is explicitly to examine the *roots* of sol-gel research.

Step 1. There was a very considerable effort among polymer chemists to synthesize new polymers as precursors for ceramics,[49] but not much is aimed at new materials. [Clabaugh et al.[50]] fixed the stoichiometry of $BaTiO_3$ by using barium titanyl oxalate as a precursor, and Hirano et al.[51] used a similar precursor with one atom each of Li and Nb to make precisely stoichiometric $LiNbO_3$ derived by hydrolysis of such precursors. Deviations from stoichiometry are connected to degradation of $LiNbO_3$ in most electro-optic applications. Hirano has also succeeded in crystallizing $LiNbO_3$ out of these solutions at temperatures as low as 300°C possibly because of the molecular structure of the precursor. Very recently Mehrotra[52] has reviewed the availability of many such bimetallic alkoxides. Except for electronic ceramics or very thin coatings applications the cost factor rules out any commercial use of complex organic precursors.

Steps 2-4. There was a considerable emphasis on the 'molecular structure" of the solution, and the understanding and control of the gelation process.[53] The ratio of hydrolyzing solution to organic polymer is one universally recognized critical parameter. [Yoldas[54]]

Step 5. Another new group of studies was concerned with making larger monoliths as aerogels usually of a closed porosity glass usually high in SiO_2. These studies extend earlier work which aimed at extracting the water out of the gel by controlling its microstructure, by the use of "drying control agents", or the use of critical point drying to avoid the formation of a meniscus. Transparent NCS several cm in diameter stable to about 800°C have been made.[53] A novel twist to this approach is the successful use of CO_2 near room temperature as the drying fluid. [Tewari et al.[55]] Using this method has made possible modest size (several cm^3) optically transparent aerogel windows with between 10 and 60% solid SiO_2. Fricke[56] has given an excellent review of the use of aerogels. Recently we have shown [Komarneni et al.[57]] the enhancement of the thermal stability of aerogels by making them diphasic (see below).

Step 6. Still another set of studies concentrates on the synthesis of electronic ceramics. It seeks to exploit the fact that from such solutions (not usually sols) the desired phase-typically a perovskite or similar structure-will crystallize at the lowest possible temperature.[58] Such low temperatures of formation of capacitor and active elements could be of great significance in incorporating SSG processing into the semiconductor industry (see below).

Step 7. In the glass area, part of the studies have been concerned with optimizing the process for making ultrahomogeneous and ultra-pure glass with the minimum content of 3d ions.[59, 60] By far the most significant work here is the application of the sol-gel process in glass fiber manufacture [MacChesney[61]].

Another set of papers is concerned with phase separation in the xerogels of compositions where the glass may or may not be known to phase separates.[61]

Transparent Ceramics

Recently by combining sol-gel precursors with the use of microwave processing we have been able to make transparent hydroxyapatite ceramics in a matter of minutes.[62] In a further refinement we have shown how only by the use of aerogel precursors can one make mullite transparent via the microwave route.[63]

THE NEW RESEARCH DIRECTION FOR SSG RESEARCH-MAXIMALLY HETEROGENEOUS MATERIALS OR "NANOCOMPOSITES"

As will have become clear from this review the principal goal and advantage of SSG processing has been to make maximally *homogeneous* glasses and ceramics, and to make them at much lower temperatures and reaction times. This ultrahomogeneity was, of course, my own goal when I initiated the work in 1948 but remains-today-the goal of virtually every investigator in the field. In 1982, however I announced the turning the *goal* of our work to using the SSG process to make *not ultrahomogeneous* but "maximally" *heterogeneous* materials[64 a,b,c]. "Maximally heterogeneous," here, can be quantified by the surface area of the contact between the two (or more) phases involved. To attain such heterogeneity we need a perfect dispersion and mixing of two (or more) phases, and the subdivision of the latter on an extremely fine scale. It was in this connection that I first introduced the term *nanocomposites* to describe such materials from the fact that the size of the individual phases is in the nanometer range (typically 1-10--nanometers). A nanocomposite is defined as a uniform mixture of such very small (in at least one direction) particles of one phase (either crystalline or noncrystalline) with one or more (either crystalline or noncrystalline) other phases.

We have described schematically two very different approaches to making diphasic xerogel nanocomposites, in earlier papers (see Ref. 1). The first method is based on our experience with crystal growth in gels. In that work the goal was to minimize nucleation and hence cause one (or a few) crystals to grow very large while not exceeding the solubility product of the growing phase throughout most of the gel host.

In the nanocomposite studies we reversed the nucleation/growth ratio making it nearly infinite and causing the uniform growth of 1-10 nm size "crystals" of the second phase throughout the present gel. By using the appropriate solutions Roy and Roy[65] and Hoffman Komarneni and Roy[66] were able to make nanocomposites of the following combination of materials:

NCS Host + Crystalline Dispersoid				NCS Host + NCS Dispersoid	
SiO_2	+ AgCl	Al_2O_3	+ Ni	SiO_2	+ $(AlO.OH)_n$
	+CdS		+ Pt		+ $AlPO_4$
	+ $BaSO_4$		+ Cu		+ $(NdO.OH)_n$
	+ $PbCrO_4$				
	+ Cu	ZrO_2	+ Cu		
	+ Ni		+ Pt		

These "0-3" (see Newnham*, 69) AgCl-SiO_2 composite gels can be made into photochromic solids or coatings simply by drying at 50°-100°C; and can also be used as precursors for melting to photochromic glasses of any suitable composition. The metal-containing composites were made by reducing the second phase which was typically a hydroxide of a transition metal, in forming gas at 300°- 500°C. Diphasic Ni-Al_2O_3 xerogels formed by this process can be sintered into hard and tough nanocomposites with hardnesses of 20 Gpa and fracture toughness near 5-10 MPa/m.[67]

The second method for making nanocomposites is the most general one widely applicable to all ceramics. It requires simply the vigorous mixing of two separate sols, assuring that no rapid flocculation of one by the other occurs and then gelling and processing. By this method some remarkably useful nanocomposites described in the next section have been produced. The SG technique is here essential to its success since by no other means can we assure the homogeneous mixing of phases on such a small scale.

SEEDING AND EPITAXY IN CERAMIC REACTIONS: THE NEW SSG RESEARCH AREA

From our very earliest work in 1948 in the systems Al2O3-H2O, Li2O-Al2O3-SiO2-H2O and thereafter, we practiced the use of adding crystallographic seeds to nucleate possibly stable or desirable phases. These seeds were tens of microns in size and were shown to be critically effective in hydrothermal experiments. In the nanocomposites, we first proved the remarkable fact that epitaxial effects dominate even in dry heating.

---

*The systematic terminology we use for describing the connectivity of all composites is due to Newnham.[69]

From a phase rule viewpoint two sols, one with a quartz solid phase and one with the usual noncrystalline SiO_2 (present in Ludox) when mixed and gelled and desiccated make a *structurally diphasic xerogel.* We first showed how such ultraheterogeneous materials behave in an extraordinary manner with respect to their reaction temperatures and the microstructures of the product. Figure 6 shows the DTA patterns of several xerogels comparing a single phase Al_2O_3- [NCS] gel with a series of diphasic Al_2O_3-[NCS] + ≈1% Al_2O_3-(corundum) gel.[70]

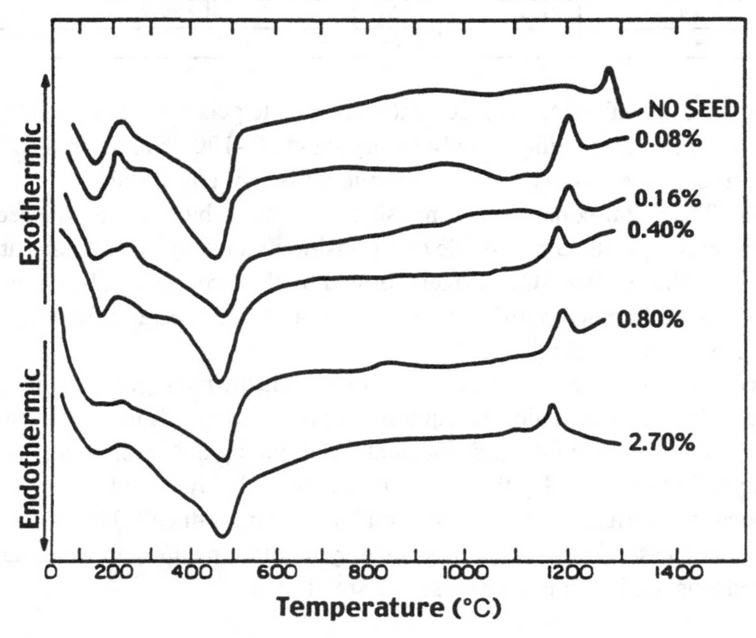

Figure 6: The surprising finding that even in the solid state epitaxial substrates (seeds of the specific structure ONLY) can very substantially influence the kinetics of fine ceramic powder crystallization. Shown above is the case of Al_2O_3 with the 150°C lowering of the θ to α transition.

It will be seen that even during the short time of the rapid heating (20°C/min) of a DTA apparatus the transition temperature and crystallization temperature of α-alumina has been lowered by 100°- 150°C. By using a series of crystalline second phases we first conclusively established that this lowering of the crystallization temperature is a true epitaxial seeding effect, i.e. catalysis of the solid state growth of a phase by provision of large numbers of nuclei of the stable (in this case) phase. Thus adding TiO_2 or SiO_2 sols as second phases have no effect on an Al_2O_3

sol. Obversely TiO_2-[rutile] sol added as seeds lowers the temperature by over 200°C [70] of rutile crystallization from TiO_2-[NCS] sols. Equally remarkable and valuable is the effect of diphasic sol-gel techniques on microstructure development.[71] Figure 7(top) shows the microstructures developed in a typical thin alumina gel fragment (of the abrasive grain size) in pure Al_2O_3; 7(bottom) shows the same parent sol phase mixed with 1 % of the Al_2O_3-[corundum] sol. The grain size is seen to drop from 10-20 μm to <1 μm.[68] Furthermore, we have also recently demonstrated that the epitactic seeding works even in very complex compositions and structures such as $NaZr_2P_3O_{12}$[72] with a marked influence on both sintering and microstructure. The most spectacular case is that of the crystallization of albite glass, which normally cannot be achieved by months of annealing close to its melting point, is completed in a day by seeding.[73]

Figure 7: The radical change of the microstructure of Al_2O_3 gel crystallization. Top: without α- Al_2O_3 seeds; bottom: with 1% such added.

COMPOSITIONALLY AND "DOUBLY" DIPHASIC GELS — REACTIVE SINTERING IN NANOCOMPOSITES

Using mullite ($3Al_2O_3.2SiO_2$) as the example we have compared the sintering behavior of a single phase xerogel (i.e. one derived from a homogeneous solution of $Al(NO_3)_3$ and SiO_2-[NCS] with a second xerogel made from two separate sols[74] and which we have shown by MASNMR[28] has separate alumina and silica phases. Of course, there is a considerable improvement in density obtained by using a diphasic gel, due to the heat of reaction which drives the sintering. But one can go further.

When the compositionally diphasic mullite gel (above) has a sol added to it with a second *structurally* different phase-i.e. *crystalline* mullite-and sintered side by side a remarkable change in *morphology* of the final crystals occurs. The equant grains of mullite are replaced by 10:1 aspect ratio mullite needles, while the greater density attained is not affected. Thus judicious mixing of structural and compositional diphasicity (including crystallographic seeding) appears to provide a most interesting and wholly novel process control mechanism for enhancing density, refining microstructure and controlling morphology.

The most informative examples which we have studied in detail are $ZrSiO_4$ and $ThSiO_4$. Figure 8 illustrates beautifully the effects on *temperature* of reaction of the various combinations of structural and compositional diphasicity.[75] The case of $ThSiO_4$ forming huttonite or thorite[76] even at 1350°C by merely adding 1% of crystalline second phase proves the enormous potential for ceramic processing in the future, by directing the phase which forms.

STRUCTURALLY \ COMPOSITIONALLY	Nanocomposite	Homogeneous
Homogeneous	1175°C	1350°C
Nanocomposite	1075°C	1100°C

Figure 8: The role of compositionally diphasic gels in lowering reaction temperatures is compared to the structurally diphasic gels, and to that where it is both structurally (seeded) and compositionally diphasic.

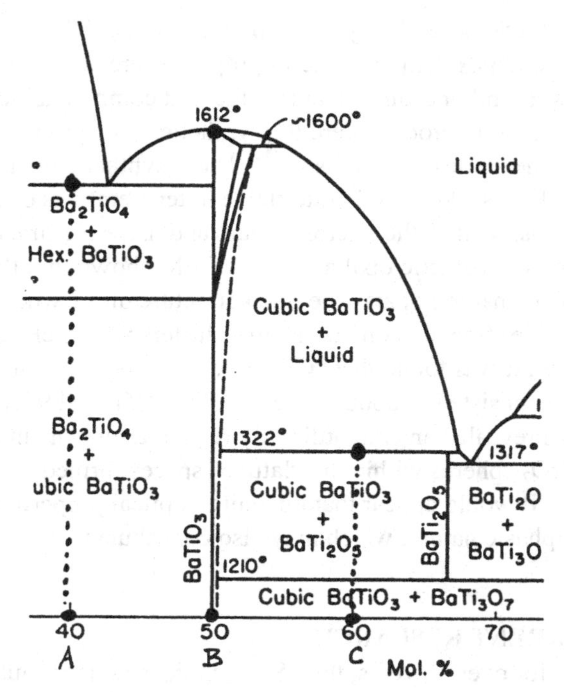

Figure 9: Section of the phase diagram for the system BaO-TiO_2. The universal potential of diphasic mixtures of any desired composition is illustrated above. One can make a final $BaTiO_3$ product by starting with a single phase gel of composition B (above) or of an equimolar mix of gels at the A and C compositions. B will (at equilibrium) develop no liquid till 1612°C. But, C will form eutectic liquids no higher than 1322°C, and intimate mixtures with "A" is thermodynamically required to form metastable liquids much lower. Hence, in ALL ceramic systems, diphasic starting materials make possible very low temperature liquid phase sintering due to metastable melting.

NATURAL NANOCOMPOSITES: LOW-TEMPERATURE CHEMICALLY BONDED CERAMICS

Nature abounds in micro- and nanocomposite structures to achieve toughness typically by combining a ceramic and a polymer. Examples include human bones and teeth where the 1000Å hydroxyapatite fibrils are topotactically related to the collagen phase. Another example is the 3:3 composite of single crystal $CaCO_3$ and a different polymer in the very hard sea-urchin spines and in corals, from which we have made "synthetic bones" very close to human bones.[77] Other gel-derived room temperature ceramics with very useful properties are also known. Furthermore, it has recently become clear that the first commercial tough and hard ceramics made by the SSG process were the synthetic opals produced continuously since the sixties by P. Gilson.[78] These synthetic opals are hard and tough ceramics made at <100°C, and quite stable to temperatures over 400°C.[76] We have recently examined[79] the microstructure and nanostructure and properties of Gilsonite synthetic opal and in the SEM shown that they are indistinguishable from natural opals. The microstructure of the wholly inorganic opal with its (probable) fractal geometry shows clusters within clusters of small SiO_2 balls. However, it was found that "Gilsonite" was in fact a true nanocomposite and consisted of about 5% crystalline ZrO_2 and 95% ncs-SiO_2. Moreover, the TEM revealed an extraordinarily regular arrangement of the much smaller (20 nm) ZrO_2 spheres within the "lattice" spaces formed by the fcc packed larger balls. It would appear that gilsonite's optical properties are directly derived from its diphasic nature, which may also contribute to its strength and toughness.

A NOTE ON CURRENT RESEARCH

As it has been for over 40 years, the SSG technique is used routinely in the Penn State Materials Research Laboratory and other departments for a wide variety of problems. The bibliography appended to the references gives the best bird's-eye view of our group's work in the field, since 1990.

CONCLUSIONS

The use of the solution sol-gel process to make ultrahomogeneous glasses and ceramics from the common oxides, using first organic, and later principally inorganic, precursors has developed from its beginnings nearly 50 years ago in the author's laboratories into a universal process for this purpose. Major technologies of modest size, utilizing the solution-sol-gel route have been widely practiced by companies all over the world for about 20 years to make nuclear fuel pellets, ceramic fibers, thin (<lμ) coatings, abrasive grain and recently, optical fiber pre-forms. Any payoff from the recent 'fashion' in sol-gel research is likely to be in new electronic ceramic materials where thin layers are adequate, where cost is no bar to the use of expensive precursors, and where chemical purity and stoichiometry-control command a premium. Aerogels are well positioned to offer a new bulk application in insulation if cost factors can be managed.

The radical redirection of SSG research in the author's laboratory towards ultraheterogeneity has opened up different vistas for lowering ceramic reaction temperatures and controlling microstructures by utilizing solid state epitaxy and topotaxy-both unexpected phenomena in 'dry' typical ceramic reactions.

ACKNOWLEDGMENTS

The earliest work was enormously helped by several persons including my mentor, Professor E. F. Osborn, and my colleagues D. M. Roy, G. Ervin and R. C. DeVries; more recent years work has been done in collaboration with Prof. S. Komarneni.

The first sol-gel research in 1948 was supported by the Office of Naval Research and the US. Army Signal Corps (1953). The second generation diphasic sol-gel research in the laboratory was supported by the AFOSR under Contract F49620-85-C-0069, and the Ceramics and Electronic Materials Program, the National Science Foundation under Grant DMR-8507912. I wish especially to acknowledge the support of my colleague S. Komarneni in organizing this symposium.

REFERENCES

[1]R. Roy, "Ceramics via the SSG Route," *Science* **238**:1664-1669 (1987)

[2]R. K. Iler, "The Colloid Chemistry of Silica and Silicates," Cornell University Press, Ithaca, NY, 1955.

[3]C.J. Brinker and G.W. Scherrer, "Sol-Gel Science: They Physics and Chemistry of Sol-Gel Processing," Academic Press, New York, 1990.

[4]T.E.Wood and H. Dislich, "An Abbreviated History of Sol-Gel Technology" pp. 3-23, Sol-gel Science and Technology, 1996.

[5]E. M. Levin, C. R. Robbins and H. F. MeMurdie, "Phase Diagrams for Ceramists," Am. Cer. Soc., Columbus GA, 1964.

[6]R. Roy and E. F. Osborn, "The System Al_2O_3-SiO_2-H_2O," *Am. Mineral.* **39**:853-885 (1954).

[7]R. Zsigmondy, "Zur Erkenntnis der Kolloide," Fischer, Jena, 1905.

[8]T. Svedberg, "Herstellung Kolloider Losungen," Academic Press, Dresden, 1922.

[9]H. Freundlich, "Colloid and Capillary Chemistry," Translated by H.S. Hatfield, E.F. Dutton, New York, 1922.

[10]Weiser and W.O. Milligan "Advances in Colloid Science," Vol. 1, Ed. E. Kraemer, Interscience, New York, 1942.

[11]W. Patrick, "Silicon Gel and Process of Making Same," U.S. Patent 1,297,724 (1922); "Catalytic and Adsorbent Gel," U.S. Patent 1,696,644 (1928).

[12]S.S. Kistler, Inorganic Aerogel Compositions, U.S. 2,188,007 (1940) and *Nature* **127**:711 (1931).

[13]A.B. Searle, "The Chemistry and Physics of Clays," Ernest Benn Ltd., London, 1933.

[14]RS. Ewell and H. Insley, *J. Res. NBS* **15**:173 (1935).

[15]S.P. Raychaudhuri and W.B. Dattu, *J. Phys. Chem.* **49**:21 (1945).

[16]F. de Korosy, *Am. Ceram. Soc, Bull.* **20**:162 (1941).

[17i]D.M. Roy and R. Roy, "An Experimental Study of the Formation and Properties of Synthetic Serpentines and Related Layer Silicate Minerals," *Am. Mineral.* **39**:957-975 (1954).

[17ii]D.M. Roy and R. Roy, *Am. Mineral.* **40**:147 (1955).

[18]DeVries, R. Roy and E.F. Osborn, *Trans. Brit. Ceram. Soc.* **53**:525 (1954).

[19]D.E. Rase and R. Roy, *J. Am. Ceram. Soc.* **38**:389 (1955).

[20i]R.C. DeVries and R. Roy, "Phase Equilibria in the System $BaTiO_3$-$CaTiO_3$," *J. Am. Ceram. Soc.* **38**, 142-146 (1955).

[20ii]R.C. DeVries and R. Roy, "Phase Equilibria in the System $CaO-TiO_2-SiO_2$," *J. Am. Ceram. Soc.* **38**, 158-171 (1955).

[21]R. Roy, "Aides in Hydrothermal Experimentation: II. Methods of Making Mixtures for Both 'Dry' and 'Wet' Phase Equilibrium Studies," *J. Am. Ceram. Soc.* **39**:145-6 (1956).

[22]W.C. Luth and C.O. Ingamells, *Am. Mineral.* **50**:255 (1965).

[23]D.L. Hamilton and C.M.B. Henderson, *Min. Mag.* **36**:832 (1968).

[24]E.F. Heald, J.R. Reeher and D. R. Herrington, *Am. Mineral.* **54**:317 (1969).

[25]A.D. Edgar "Experimental Petrology: Basic Principles and Techniques," Clarendon Press, Oxford, 1973.

[26i]G.J. McCarthy, R. Roy and J.M. McKay, "Preliminary Study of Low Temperature 'Glass' Fabrication From Non-Crystalline Silicas," *J. Am. Ceram. Soc.* **54**:637-638 (1971); *ibid* **54**:639 (1971).

[26ii]G.J. McCarthy, R. Roy and J.M. McKay, "Gel Route to Homogeneous Glass Preparation: II. Gelling and Desiccation," *J. Am. Ceram. Soc.* **54**:639-640 (1971).

[27]M.F. Bechtold and O.E. Snyder, U.S. Patent 2,574,902 (DuPont) 1951.

[28]R.K. Iler, "The Chemistry of Silica, Solubility. Polymerization, Colloid and Surface Properties, and Biochemistry," Wiley, New York, 1979.

[29]E. Matijevic, *Acc. Chem. Res.* **14**:22 (1981); *Ann. Rev. Mat. Sci.* **15**:483 (1985).

[30]S. Komarneni, R. Roy, C.A. Fyfe, G.J. Kennedy and H. Strobl, "Solid State ^{27}Al and ^{29}Si Magic Angle Spinning NMR of Aluminosilicate Gels," *J. Am. Ceram. Soc.* **69** (31):C42-C44 (1986).

[31i]F.A. Mumpton and R. Roy, "Low Temperature Equilibria Among ZrO_2, ThO_2 and VO_2," *J. Am. Ceram. Soc.* **43**:234 (1960).

[31ii]R. G. Wymer, "Status Report from the U.S.; Sol-gel Work on Ceramic Fuel Production," *Sol Gel Processes for Ceramic Nuclear Fuels*, IAEA, Panel Proceeding Series, Vienna, 1968.

[32]J.P. McBride, "Oak Ridge National Laboratory Report," ORNL-3874 (1966); M.E.A. Hermans, *Sci. Ceram.* **5**:523 (1970).

[33]Patents on alumina fibers: U.S. Patents 3,322,865 (1964); 3,503,765 (1966); 3,808,015 (1967); 3,705,223 (1968); British Patents 1,360,197, 1,360,199, 1,402,544, 1,231,385.

[34]Patents on mullite-rich fibers. U.S. Patents 4,159,205 (1978), 4,277,269 (1979) and 4,047,965 (1977).

[35]Patents on other oxide fibers. U.S. Patents 3,860,529 (1968), 3,385,915 (1967), 3,793,041 (1974), 3,709,706 (1973), 3,909,278 (1975) and 4,125,406 (1978).

[36]J.D. Birchall, *Brit. Ceram. Trans. and J.* **83**,158 (1984); "Handbook of Composites," Vol. 1, Ed. W. Watt and B.V. Perov, p. 115 (1985).

[37]S. Yajima, K. Okamura, J. Hayashi and M. Omori, "Synthesis of Continuous SiC Fibers with High Tensile Strength," *J. Am. Cer. Soc.* **59**:324 (1976).

[38]S. Yajima. Chem. Lett. **9**:931 (1975).

[39]W. Geffcken. *Glastechn. Ber.* **24**:143 (1951); H. Dislich et al. FRG Patents 1,494,872 (1965); 1,941,191 (1969); U.S. Patent 4,229,491.

[40]H. Dislich and D. Hinz, *J. Noncryst. Sol.* **48**:11 (1982).

[41]H. Dislich and E. Hussmann, *Thin Solid Films* **77**:129 (1981).

[42]B.E. Yoldas, U.S. Patent 4,361,598 (1983).

[43]Ian Thomas, *Mat. Res. Soc. Annual Meeting Abstracts* (1982).

[44]I. Pope and D. Harrison, *ibid* (1980).

[45]M.A. Leitheiser and H.G. Sowman, "Non-Fused Aluminum Oxide Based Abrasive Mineral," U.S. Patent 4,314,827.

[46]R. Roy and D. Agrawal, J.P. Cheng and M.Mathis, "Microwave Processing. Triumph of Applications-Driven Science in WC-Composites and Ferroic Titanates," *Proc. First Int'l. Symp. on Microwave Processing*, Am. Ceram. Soc., Westerville, OH, 1997.

[47]The most recent of these are: L.L. Hench and D.R Ulrich, Eds., Ultrastructure Processing of Ceramics, Glasses, and Composites (Wiley, New York, 1984); C.J. Brinker, D.E. Clark, and D.R. Ulrich, EDs., Better Ceramics Through Chemistry (Elsevier, New York, 1984); L.L. Hench and D.R. Ulrich, Eds., UltraProcessing of Ceramics, Glasses, and Composites (Wiley, New York, 1986).

[48]B.E. Yoldas, *Applied Optics* **21**:2960 (1982); B.E. Yoldas and T.W. O'Keeffe, *Applied Optics* **18**:3133 (1979); Y. Suwa, Y. Sugimoto and S. Naka, *J. Japan Soc. Powder and Powder Metallurgy* **25**:164 (1978); Y. Suwa, Y. Kato, S. Hirano and S. Naka, *J. Japan Soc. Powder and Powder Metallurgy* **28**:28 (1981); D. Seyferth and C.H. Wiseman, "Processing of Ceramics, Glasses and Composites," Eds. L.L. Hench and D.R. West, *ibid*, p. 235.

[49]C.J. Brinker et al., *J. Non-Crystl. Solids* **48**:47 (1982) ; B.E. Yoldas, *ibid* **63**:145 (1984); C.J. Brinker and G.W. Scherer, ibid **70**:301 (1985); C.J. Brinker et al., *ibid* **71**:171 (1975); C.J. Brinker et al., *ibid* **63**:45 (1984) ; M. Prassas and L.L. Hench "Processing of Ceramics, Glasses, and Composites," L.L. Hench and D.R. Ulrich, Eds., Wiley, New York, 1984.

[50]W.S. Clabaugh, E.M. Swiggard and R. Gilchrist. *J. Res. NBS* **56**:289 (1956).

[51]S. Hirano, personal communication, 1985.

[52]R. C. Mehrotra, *Proc. Ind. Acad. Sci.* **61**:253 (1995).

[53]J. Zarzycki, M. Prassas and J. Phalippou, *J. Mat. Sci.* **17**:3371 (1982) M. Prassas et al., *J. Non-Cryst. Solids* **48**, 79 (1982); F. Pancrazi et al., *ibid* **63**:81 (1984); L.L. Hench, M. Prassas and J. Phalippou, *ibid* **53**, 183 (1982); S.P. Mukhejee, *ibid* **42**:477 (1980).

[54]B.E. Yoldas, *J. Am. Ceram. Soc.* **65**:387 (1982).

[55]P.H. Tewari, A.J. Hunt and K.D. Lofftus, *Mat. Lett.* **3**:363 (1985).

[56]R. Fricke and A. Emmerlin, *J. Am. Ceram. Soc.* **75**(8):2027 (1992).

[57]S. Komarneni, R. Roy, U. Selvaraj, P.B. Malla and E. Breval, "Nanocomposite Aerogels: The SiO_2-Al_2O_3 System," *J. Mater. Res.* **8**(12):3163-3183 (1993).

[58]David Payne, Paper presented at Ohio Ceramic Society Meeting, Cleveland, OH (Spring 1985).

[59]E.M. Rabinovich et al., *J. Am. Ceram. Soc.* **66**:683 (1983) ;

[60]D.W. Johnson, Jr. et al., *ibid* **66**:688 (1983).

[61]J.B. MacChesney, D.W. Johnson, Jr., S. Bhandarkar, M.P. Bohrer, J.W. Fleming, E.M. Monberg and D.J. Trevor, "Optical Fibers Using Sol-Gel Silica Overcladding Tubes," *Electronics Letters* **33**(8):1573-4 (1997).

[62]Y. Fang, D.K.Agrawal, R. Roy, D.M. Roy, "Fabrication of Transparent Hydroxyapatite Ceramics by Ambient-Present Sintering," *Mat. Lett.* **23**:147 (1995).

[63]Y. Fang, D.K. Agrawal, R. Roy, and D. Roy "Transparent Mullite Ceramics from Diphasic Aerogels by Microwave and Conventional Processings," *Mat. Lett.* **28**:11-15 (1996)

[64i]"New Hybrid Materials Made by Sol-Gel Technique," *Bull.Am. Cer.Soc.*, **61**:374 (1982).

[64ii]"Ceramics from Solutions: Retrospect and Prospect," p. 370, *Mat. Res. Soc., Annual Meeting Abstracts*, 1982.

[64iii]"New Metal-Ceramic Hybrid Xerogels," p. 377 *Mat. Res. Soc., Annual Meeting Abstracts*, 1982.

[65]R.A. Roy and R. Roy, "Diphasic Xerogels: I. Ceramic-Metal Composites," *Mat. Res. Bull.* **19**:169-177 (1984).

[66i]D. Hoffman, S. Komarneni and R. Roy, "Preparation of a Diphasic Photosensitive Xerogel," p. 429-442, *J. Mat. Sci. Lett.* **3**:439 (1984).

[66ii]D. Hoffman, R. Roy and S. Komarneni, "Diphasic Ceramic Composites Via a Sol-Gel Method," *Mat. Lett.* **2**:245 (1984).

[66iii]D. Hoffman, R. Roy and S. Komarneni, "Diphasic Xerogels, A New Class of Materials: Phases in the System Al_2O_3-SiO_2," *J. Am. Ceram. Soc.* **67**:468 (1984).

[67]E. Breval, G. Dodds and C.G. Pantano, *Mat. Res. Bull.* **20** (1985).

[68i]Y. Suwa, R. Roy and S. Komarneni, "Lowering Crystallization Temperatures by Seeding in Structurally Diphsic Al2O3-MgO Xerogles," *J. Am. Ceram. Soc.* **68**:C238-C240 (1985).

[68ii]Y. Suwa, S. Komarneni and R. Roy, "Solid State Epitzxy Demonstrated by Thermal Reactions of Structurally Diphasic Xerogels: The System Al_2O_3," *J. Mat. Sci. Lett.* **5**:21-24 (1986).

[69]R. E. Newnham, *J. Mater. Educ.* **7**:601 (1985).

[70]Y. Suwa, R. Roy and S. Komarneni, "Epitaxial Effect of Seeding on Crystallization and Densification of Diphasic Al_2O_3-MgOXerogels," *Sintering* Vol. 1, pp. 170-175 (1989).

[71]W.A. Yarborough and R. Roy, "Microstructural Evolution in Sintering of AlOOH Gels," *J. Materials Res.* **2**(4):494 (1987).

[72]S. Limaye, D.K. Agrawal and R. Roy, "Synthesis, Sintering and Thermal Expansion of $Ca_{1-x}Sr_xZr_4P6O_{2\,(4)}$ — An Ultra Low Thermal Expansion Ceramic System," *J. Mat. Sci.* **26**:93 (1990).

[73]U. Selvaraj, C.L. Liu, S. Komarneni and R. Roy, "Epitaxial Crystallization of Seeded Albite Glass," *J. Am. Ceram. Soc.* **74**(6):1378 (1991).

[74]Y. Suwa, R. Roy and S. Komarneni, *J. Am. Ceram. Soc.* **68**(9):C-238 (1985).

[75]G. Vilmin, S. Komarneni and R. Roy, *J. Mat. Sci.* **22**:3556 (1987).

[76]G. Vilmin, S. Komarneni and R. Roy, "Crystallization of $ThSiO_4$ from Structurally and/or Compositionally Diphasic Gels," *J. Mater. Res.* **2**(4):489 (1987).

[77]D.M. Roy, U.S. Patent 3,929,971 (1975).

[78]Now available from the Nakazumi Crystal Laboratory, 3-1-304 Sugahara-cho, Ikeda-Shi, Osaka-fu 563, Japan.

[79]T.C. Simonton, R. Roy, S. Komarneni and E. Breval, "Microstructure and Mechanical Properties of Synthetic Opal: A Chemically Bonded Ceramic," *J. Mat. Res.* **1**:656-674 (1986).

MOST RECENT S-G WORK

Penn State MRL Recent (1990-1998) Papers Using the Sol-Gel Process

D.K. Agrawal, and J.H. Adair, "Low-Temperature Sol-Gel Synthesis of $NaZr_2P_3O_{12}$," *J. Am. Ceram. Soc.* **73**[7]:2153-55 (1990).

E. Breval, C.G. Pantano, "Sol-Gel Prepared Ni-Alumina Composite Materials, Part II Structure and Hot-Pressing Temperature," *J. Mater. Sci.* **27**:5463-69 (1992).

E. Breval, Z. Deng, S. Chiou, C.G. Pantano, "Sol-Gel Prepared Ni-Alumina Composite Materials, Part I Microstructure and Mechanical Properties," *J. Mater. Sci.* **27**:1464-68 (1992).

Y. Fang, D. K. Agrawal, D.M. Roy, R. Roy, "Fabrication of Transparent Hydroxyapatite Ceramics by Ambient-Pressure Sintering," *Mater. Lett.* **23**:147-151 (1995).

Y. Fang, R. Roy, D.K. Agrawal, and D.M. Roy, "Transparent Mullite Ceramics from Diphasic Aerogels by Microwave and Conventional Processings," submitted to *Mater. Lett.*, Y. Fang, J. Cheng, D.K. Agrawal, D.M. Roy and R. Roy, "Microwave Processing of Diphasic Aluminosilicate Gel," to be published.

Y. Fang, Q. Li, D.K. Agrawal, D.M. Roy and R. Roy, "Microwave Sintering of Mullite Ceramics," presented at *MRS Spring Meeting*, (paper No. L9.8) Materials Research Society, San Francisco, CA (1991).

Y. Fang, J. Cheng, R. Roy, D.M. Roy and D.K. Agrawal, "Fabrication of Transformation Toughened Ceramics by Microwave Processing," pp. 211-213, Proceedings, *29th Microwave Power Symposium, International Microwave Power Institute*, Chicago, IL (1994).

H. Kido, S. Komarneni and R. Roy, "Preparation of $La_2Zr_2O_7$ by Sol-Gel Route," *J. Am. Ceram. Soc.* **74**[2]:422-24 (1991).

S. Komarneni and R. Roy, "Mullite Derived from Diphasic Nanocomposite Gels," *Ceramic Transactions* Vol. 6, *Mullite and Mullite Matrix Composites*, edited by S. Somiya, R.F. Davis, J.A. Pask, pp. 209-219, Am. Ceram. Soc., Westerville, Ohio (1990).

S. Komarneni, R. Roy, U. Selvaraj, P.B. Malla and E. Breval, "Nanocomposite Aerogels: The SiO_2-Al_2O_3 System," *J. Mater. Res.* **8**[12]:3163-67 (1993).

K. Kushida, K.R. Udayakumar, S.B. Kropanidhi and L.E. Cross, "Origin of Orientation in Sol-Gel-Derived Lead Titanate Films," *J. Am. Ceram. Soc.* **76** [51]:1345-48 (1993).

C. Liu, H. Zhang, S. Komarneni and C. Pantano, *J. Sol-Gel Science and Technology* **1**:141 (1994).

C. Liu and S. Komarneni, "Carbon-Silica Xerogel and Aerogel Composites," *J. Porous Mater.* **l**[l]:75-94 (1995).

J. Majling, S. Komarneni and V.S. Fajnor, "Mercury Porosimeter as a Means to Measure Mechanical Properties of Aerogels," *J. Porous Mater.* **1**[11]:91-95 (1995).

P.B. Malla, S. Komarneni, H. Taguch and H. Kido, "Probing the Nature and the Structure of Pores in Silica Xerogels by Water Sorption: The Tetramethyl Orthosilicate-Hydrogen Chloride/Fluoride System," *J. Am. Ceram. Soc.* **74**[12]:2988-95 (1991).

L. Pach, R. Roy and S. Komarneni, "Nucleation of Alumina in Boehmite Gel," *J. Mater. Res.* **5**[2]:278-285 (1990).

A.V. Prasadarao, U. Selvaraj, S. Komarneni, A.S. Bhalla and R. Roy, "Enhanced Densification by Seeding of Sol-Gel-Derived AluminumTitanate," *J. Am. Ceram. Soc.* **75**[61]:1529-33 (1992).

P. Ravindranathan, S. Komarneni, A.S. Bhalla and R. Roy, "Synthesis and Dielectric Properties of Solution Sol-Gel-Derived $0.9Pb(Mg_{1/3}Nb_{2/3})O_3$-$0.lPbTiO_3$ Ceramics," *J. Am. Ceram. Soc.* **74**[12]:2996-99 (1991).

P. Ravindranathan, S. Komarneni and R. Roy, "Solid-State Epitaxial Effects in Structurally Xerogel of $Pb(Mg_{1/3}Nb_{2/3})O_3$," *J. Am. Ceram. Soc.*, **73**[4]:1024-25 (1990).

P. Ravindranathan, S. Komarneni and R. Roy, "Studies on Polymeric $Pb(Mg_{1/3}Nb_{2/3})O_3$ Gels," *Mater. Lett.* **8**[5]:161-64 (1990).

P. Ravindranatlian, S. Komarneni, A.S. Bhalla and R. Roy, "Dielectric Properties of Sol-Gel Derived Lead Magnesium Niobate Ceramics," *Ferroelectrics Lett.*, **11**:137-144 (1990).

C. Rutiser, S. Komarneni and R. Roy, "Composite Aerogels of Silica and Minerals of Different Morphologies," *Mater. Lett.* **19**:221-24 (1994).

C. Rutiser, S. Komarneni and R. Roy, "Composite Aerogels: Mechanical, Thermal Stability and Porosity Properties," submitted to *Ceramics International.*

C. Rutiser, "Thermal Stability, Porosity and Mechanical Properties of Composite Silica Aerogels," M.S. Thesis, The Pennsylvania State University, University Park, PA, 1995.

U. Selvaraj, S. Komarneni and R. Roy, "Synthesis of Glass-Like Cordierite Front Metal Alkoxides and Characterization by ^{27}Al and ^{29}Si MASNMR," *J. Am. Ceram. Soc.* **73**[12]:3663-69 (1990).

U. Selvaraj, A.V Prasadarao, S. Komarneni and R. Roy, "Sol-Gel $SrTiO_3$ Thin Films from Chemically Modified Alkoxide Presursors," *Mater. Lett.* **12**:306-10 (1991).

U. Selvaraj, A.V Prasadarao, S. Komarneni and R. Roy, "Sol-Gel Fabrication of Epitaxial and Oriented TiO_2 Thin Films," *J. Am. Ceramic. Soc.* **75**[5]:1167-70 (1992).

U. Selvaraj, K. Brooks, A.V Prasadarao, S. Komarneni, R. Roy and L.E. Cross, "Sol-Gel Fabrication of $Pb(Zr_{0.25}Ti_{0.48})O_3$ Thin Films Using Lead Acetylacetonate as the Lead Source," *J. Am. Ceram. Soc.* **76**[6]:1441-44 (1993).

D. Ravichandran, R. Roy, W.B. White and S. Erdei, "Synthesis and Characterization of Sol-Gel Derived Hexaaluminates Phosphors," *J. Mater. Res* **12**:819-824, 1997.

D. Ravichandran, R. Roy and W.B. White, "Low Temperature Synthesis and Particle Size Control in Yttrium Based Phospohors," *J. Soc. Info. Display* **5/2**:107 (1997).

D. Ravichandran, K. Yamakawa, A.S. Bhalla and R. Roy, "Alkoxide Derived $SrBi_2Ta_2O_9$ Phase Pure Powder and Thin-Films," *Sol-Gel Sci. & Tech.* **9**:95-101 (1997).

D. Ravichandran, R. Roy, B. Jin and A.S. Bhalla, "A Study of Sol-Gel Derived $Sr_2(AlTa)O_6$ by Raman Spectroscopy," *J. Mater. Sci. Lett.* **15:**805-806 (1996).

D. Ravichandran, B. Jin, R. Roy and A.S. Bhalla, "Raman Spectroscopy of Sol-Gel Derived $BaMg_{1/3}Ta_{2/3}$ Perovskites," *J. Mater. Lett.* **25:**257-259 (1995).

D. Ravichandran, A,.S. Bhalla and R. Roy, "Sol-Gel Derived $Pb(Sc_{0.5}Ta_{0.5})O_3$ Powder," *J. Mater. Lett.* **25**:161-63 (1995).

D. Ravichandran, R. Roy, R. Meyer, Jr., R. Guo, A.S. Bhalla and L.E. Cross, "Sol-Gel Derived $BaMg_{1/3}Ta_{2/3}$ Thin Films Substrates for High Temperature Superconductors," *Mater. Res. Bullt.* **31**:817-25 (1996).

D. Ravichandran, R. Roy, R. Guo, A.S. Bhalla and L.E. Cross, "Fabrication Of $Y_3Al_5O_{12}$: Eu Thin-Films and Powders for Field Emission Applications," (under communication).

D. Ravichandran, R. Roy, A.G. Chakhovskoi, C.E. Hunt, W.B. White and S. Erdei, "Low Temperature Synthesis of SAT Thin Films and Powder for High Tc Oxide Substrates," *J. Luminescence* **71**:291-297 (1997).

D. Ravichandran, R. Roy, A.G. Chakhovskoi, W.B. White and C.E. Hunt, "Alkoxide Derived Field Emission Display Phosphors," *J. Soc. Info. Display* (in press 1996)

D. Ravichandran, R. Roy, A.S. Bhalla and R. Guo, "Sol-Gel Derived $Sr_2Bi_2Nb_2O_9$ Phase Pure Powders and Thin-Films," *J. Am. Ceram. Soc* (in press).

HISTORY OF SOL-GEL TECHNOLOGY IN JAPAN

Sumio Sakka
Fukui University of Technology
3-6-1 Gakuen
Fukui, 910-8505, Japan

ABSTRACT

Achievements aiming at practical application of the sol-gel technique in Japan have been reviewed. It is shown that sol-gel materials of various shapes,, microstructures and properties were commercialized in response to emerging technologies. It is also shown that the most successful commercializations were made in materials unique to the sol-gel method.

INTRODUCTION

In Japan a number of ceramic scientists tried to make starting powders for sintered ceramics by precipitating from inorganic solutions in 1960's. Fine particles without coagulation were necessary to obtain well densified ceramics.

Around 1970, Mazdiyasni's group prepared dense ferroelectric ceramics from metal alkoxides [1], and Roy [2] and Dislich [3] showed that glasses could be prepared through a gel route at relatively low temperatures. These achievements led to the development of sol-gel activities in the fields of glasses and ceramics in Japan. For instance, a paper on gel-glass was presented at the 10th International Congress on Glass [4]. The interest in the sol-gel method was greatly promoted by the start of the International Workshop on Glasses from Gels [5] and other international conferences on sol-gel method which started around 1981. Since then the sol-gel method was developed not only in academic world but also in industries.

It can be said that the sol-gel method was taken up as one of the material processing techniques since around 1970. Considering that the sol-gel method is a material processing technique, I will describe the history of Japanese sol-gel activities based on the commercialized or commercializable products. They are classified as follows:

(1) Bulk materials
(2) Coating films
(3) Fibers and particles
(4) Advanced future materials

BULK MATERIALS

Table 1 lists up examples of sol-gel bulk materials developed in Japan.

To the extent authorized under the laws of the United States of America, all copyright interests in this publication are the property of The American Ceramic Society. Any duplication, reproduction, or republication of this publication or any part thereof, without the express written consent of The American Ceramic Society or fee paid to the Copyright Clearance Center, is prohibited.

Table I. Bulk materials developed in Japan

	Year	Material	Author	Manufacturer	Commercialization
(1)	1982	Silica preform for optical fibers	Susa [6]	Hitachi	No
	1988	Large silica glass	Toki [8]	Seiko-Epson	No
(2)	1987	Mullite powder	Ismail [9]	Chichibu-Onoda Cement	Yes
	1988	Machinable ceramics	Hamasaki [10]	Mitsui Mining Material	Yes
	1989	Alumina tape (film)	Shinohara [11]	Mitsubishi Material	Yes
(3)	1992	Graded index lense	Yamane [12]	Olympus Camera	Yes
(4)	1996	Porous silica for chromatography	Nakanishi [13]	Merck (Germany)	Yes
(5)	1997	Aerogel	Yokoyama [14]	Matsushita Electric	Yes

(1) Silica preform for optical fiber and large-size silica glass.

When the sol-gel technique attracted an attention of glass scientists and engineers in 1970's, most of them made considerable efforts in applying this technique to preparation of bulk silica glass. Since optical communication using silica-based optical fibers was at its start and development, a special attention was paid to sol-gel preparation of transparent silica glass rods as optical fiber preform all over the world. In Japan, Susa et al. prepared silica glass rods for drawing of optical fiber from silicon alkoxide solution without the use of supercritical drying technique [6]. The drawn fibers showed a very low loss (1.8 dB/km at 1.6 μm). The optical silica fiber preform was not commercialized, possibly because sol-gel prepared silica glass fibers had a high optical loss when doped with GeO_2 [7].

Toki et al. of Seiko-Epson Company prepared large-size transparent silica glasses (rods larger than 50 mm in diameter and one meter in length and plates larger than 12 mm in thickness and 15 cm x 10 cm in area) were prepared by adding colloidal silica to the starting solution [8].. They are not commercialized, however, possibly because they do not have any marked advantages in quality and manufacturing cost over silica glasses fabricated in conventional methods.

(2) Mullite powder, machinable ceramics and alumina tape (film).

The sol-gel processing was applied to preparation of starting powders for sintered ceramics around 1970. It was expected that the sol-gel prepared starting materials give dense ceramics by heating at relatively low temperatures. In Japan, three kinds of sol-gel commercial products appeared in late 1980's. It should be noted that around this time, very dense mullite ceramics revived as new mullite.

Ismail et al. of Chichibu-Onoda Cement [9] showed that gel powder prepared from the $3Al_2O_3{\bullet}2SiO_2$ sol consisting of boehmite sol and silica suspension becomes mullite at 1400°C and forms a very dense ceramic having a density higher than 98 % of theoretical density at 1650°C. The mullite ceramics thus prepared are characteristic of high mechanical strength. This powder is commercialized.

The machinable mica ceramics can be cut by conventional metallic saw. Sol-gel prepared powder for machinable mica ceramics could be sintered at temperatures 100°C lower than the conventional powder compact [10]. They are commercialized in Mitsui Mining Material Company for the use as insulating material for precision machining and substrate for electronic parts .

Alumina tapes were prepared by firing green sheets formed from very viscous sols by doctor blade technique at 1330°C [11]. Dense alumina tapes thus obtained, 30-100 μm thick, are commercialized in Mitsubishi Material Company for the use as ceramics for medical instrument, microphone vibrating film and microelectronics.

(3) Graded index lenses.

Graded index lenses are transparent rods or discs made of glass or plastic, which have refractive index distribution in the diretion of radius or length. They have little aberration and used as lenses for copying machine, connection of optical fibers, camera and so on. These lenses have been developed since 1970's. Graded index glass lenses have been prepared by ion exchange in dense glass or incorporation of high refractive index oxide into porous glass. Yamane and Inami [12] applied sol-gel method to the commercial production of graded index rod lenses.

(4) Porous silica rod for chromatography.

Usually, gels are microporous and are used as filters and catalyst supports. Nakanishi [13] developed a new sol-gel method for making porous silica gels of double pore system from silicon alkoxide-water soluble organic polymer compositions. Macropores and mesopores are formed as a result of concurrent gelation and phase separation due to spinodal decomposition. In these materials, the size of macropores and that of mesopores can be well controlled independently. Macropores are used for transport of substrate and mesopores are used for interaction of substrate molecules with pore surfaces. The porous monolith rods of about 10 mm diameter amd 100 mm length can be prepared, and it is expected that in near future, these gels will be produced as chromatograph elements, replacing some of currently used silica powder compacts.

(5) Aerogel

In 1997, Matsushita Electric Company anounced commercialization of bulk aerogels [14]. They developed hydrophobic, transparent aerogel plates of 10 mm in thickness and10 cm x 10 cm in aera. The highest quality aerogel, that is, the most transparent aerogel is for the Cerenkov effect measurement in observation of high energy particles in physics. Aerogels of lower prices are for insulation.

COATING FILMS

Transparent coating films prepared by solution method were already applied to glasses in Schott of Germany in 1960's. Schottt developed and commercialized many kinds of sol-gel coating films in early 1980's. In Japan, a large number of sol-gel coating films are commercialized, as shown in Table II. Many of them have been developed in agreement of technological demands in each field.

Table II Coating films developed in Japan

	Year	Material	Author	Manufacturer	Commercialization
(1)	1982	In_2O_3 conducting film	Ogiwara [15]	Hitachi	Not known
(2)	1988	Micropatterning by stamping	Tohge [16]		No
(3)	1988	HUD combiner for windshield	Makita [17]	Central Glass	Yes
(4)	1989	Protecting film on steel	Izumi [18]	Nisshin Steel	Yes
(5)	1990	Coatings for CRT: Selectively absorbing	Itoh [19]	Toshiba	Yes
	1994	Coatings for CRT Antireflection	Hayama [20]	Matsushita Electronics	Yes
(6)	1992	Coating for automobiles: Water-repellent	Yamasaki [21]	Central Glass	Yes
			Morimoto [22]	Asahi Glass	Yes
	1996	Coating for automobiles: UV-shielding	Morimoto [23]	Asahi Glass	Yes
(7)	1996	Protecting coating for building materials	Yamada [24]	JSR (Japan Synthetic Rubber)	Yes
(8)	1996	Transparent photocatalyst	Kato [26]	Kato Manufacturing	Yes

(1) In_2O_3:Sn conducting films.

Ogiwara and Kinugawa [15] prepared In_2O_3:Sn (ITO) conducting films on soda-lime-silica glass from solutions containing acetylacetonates of indium and tin. This film is very important as transparent electrode, and many materials scientists are trying to prepare films of lower resistivity ($1x10^{-4}$ Ω•cm). The sol-gel prepared film showed specific resistivity of 5.6×10^{-4} Ω•cm , which was three times higher than that of vacuum-deposited coating film. In order to protect ITO film from the attack of alkali ions diffusing from the substrate, a thin SiO_2 layer was deposited before ITO coating. Sol-gel derived ITO electrode is not commercialized yet.

(2) Micropatterning by stamping.

Micropatterning is used for preparing optical waveguides or memory disc. Sputtering and lithography are main methods for micropatterning. In 1980's Tohge et al.[16] developed a sol-gel micropatterning method, in which SiO_2-TiO_2

sol-gel coating film is mechanically stamped with a stamper. For instance, grooves of width less than one micron can be prepared.

(3) HUD combiner for windshield of automobiles.

An HUD (Head up display) combiner is a light reflecting film deposited on the windshield of automobiles. With this system the driver can see the speed of the car without looking down on the panel board. The speed is displayed on the windshield. Makita et al. [17] developed a SiO_2-TiO_2 coating film, about 200 nm thick, deposited on the windshield as combiner. This film shows more than 20 % reflectivity, which permits seeing the displayed signs clearly, and more than 70 % transmittance, which agrees to the legal regulation concerning the visibility of the windshield. It should be noted that this triggered the commercialization of the sol-gel coating films as applied to automobile windows.

(4) Protecting film on steel.

In 1980's scientists and engineers in other fields than glasses and ceramics started to be interested in sol-gel technology. Commercialization of sol-gel protecting film on steel is one of the results. Izumi et al. [18] developed protecting (oxydation-resistant and corrosion-resistant) coating films based on SiO_2. A part of $Si(OC_2H_5)_4$ as source of SiO_2 was replaced by $CH_3Si(OC_2H_5)_3$, in order to give the coating a flexibility necessary for bending the steel sheets after coating without crack formation.

(5) Coating for CRT: Selectively absorbing.

In 1980's there was a trend of improving clear vision of color television picture by coating or surface treatment of the panel glass of CRT tube in response to increasing demand in the consumer TV market for higher picture quality. Itoh et al. of Toshiba [19] adopted the sol-gel method, in order to coat the panel glass with a selevtively absorbing coating film. SiO_2-ZrO_2 coating film containing organic pigment which absorbs the disturbing lights, achieving remarkable increase in contrast, was successfully applied to the panel glass.

Hayama et al. of Matsushita Electronic Company [20] coated the panel glass with three layers. The lower layer is conducting high refractive index layer of SiO_2-TiO_2 system with dispersed Sb-doped SnO_2 particles. The intermediate layer is low refractive index SiO_2 layer. The upper layer is low refractive index layer with uneven structure. Accordingly, the coating film gives a low-reflection and non-glare effect and antistatic effects.

(6) Coatings for automobiles: water-repellent, UV-shielding and antireflecting films.

In these years, demands from consumers have resulted in intensive attempts to donate automobile window glasses and mirrors with various new functions by coating. Originally, most coating films of this type were deposited by vapor phase depositions, such as vacuum deposition, sputtering and chemical vapor deposition, but some people tried sol-gel coating.

Water-repellent coatings consisting of SiO_2 doped with fluoroalkyltrimeth-oxysilane ($CF_3(CF_2)_7CH_2CH_2Si(OCH_3)_3$), a water-repellent fluorine compound, were applied to automobile windows and mirrors by Central [21] and Asahi [22] Glass Companies. These windows and mirrors assure clear sight on rainy days.

A UV shielding glass for automobiles was developed by sol-gel method in Asahi Glass Company [23], in order to protect passengers from exposure to UV lights. The coating film consisted of two layers: an upper layer of CeO_2-TiO_2 complex oxide absorbing ultraviolet rays and a lower layer of low refractive index. The lower layer is used to avoid unfavorable coloring of the coated window due to the selective light reflection by the upper layer. Without the lower layer, coated window looks yellowish.

(7) Protecting coating for building materials.

JSR Company (JSR: Japan Synthetic Rubber) commercialized sols for organic-inorganic hybrid thick coating films (paints) for building walls [24]. The final product of thick, transparent coating film consists of polyorganosiloxanes containing colloidal silica or alumina particles. In order to prepare the sol, methyl-triethoxysilanes are hydrolyzed and polycondensed under the polymerization catalyst effect of aluminum ethylacetoacetate ($Al(CH_3COCH_2COOC_2H_5)_3$). When the molecular weight of polyorganosiloxanes reaches a given value, acetylacetone is added to the solution, in order to suppress the further polymerization which leads to unfavorable gelation before use. When acetylaceton is added to the solution, ligand exchange occurs, changing aluminum ethylacetoacetate into aluminum acetyl-acetonato ($Al(CH_3COCH_2COCH_3)_3$) and the polymerization rate becomes much smaller. This makes the preservation durability (pot life) of the sol very long (longer than 4 months at 45 °C). The resultant sol is applied to the building walls as hard, durable coatings, thicker than 20 μm. JSR developed another type of sols for very thick coatings from acryl polymers containing alkoxysilyl groups at side chains and alkoxysilanes. The success of sol-gel coating for building walls indicates that application of sol-gel processing should not be limited to high technology materials.

(8) Transparent photocatalyst.

When TiO_2 is irradiated with UV light in aqueous solution, water is decomposed into H_2 and O_2 . It was shown that sol-gel prepared TiO_2 films are very effective in causing this photocatalyst effect [25]. In these years, the decomposition of organic molecules by photocatalyst effect of TiO_2 has been intensely studied for the purpose of deodorization in rooms and self-cleaning of building walls and car windows.[26]. Anatase phase of TiO_2 is more effective than rutile phase TiO_2.

Kato [27] developed TiO_2 coated glass containers. It was shown that trans-parent TiO_2 films applied to glass containers decompose organic odorant molecules or contaminating molecules in water.

FIBERS AND PARTICLES

Sol-gel method is suitable for preparing fibers, because fiber drawing is possible from viscous sols at low temperatures near room temperature. Many kinds of fibers were shown to be prepared by this method. Two kinds of fibers are shown in Table III. Besides these, SiC fiber named Nicalon is a very famous and important sol-gel product. Silica spheric particles are also commercialized.

Table III Fibers and particles

	Year	Material	Author	Manufacturer	Commerciali-zation
(1)	1974	Alumina fibers	Horikiri [28]	Sumitomo Chemical	Yes
(2)	1988	Silica fibers	Taneda [29]	Asahi Glass	Yes
(3)	1996	Silica spheres	Adachi [30]	Ube-Nitto Chemical	Yes

(1) Alumina fiber.
It is remarkable that heat-resistant alumina fibers doped with 15 % silica were commercialized at the early days of the present sol-gel method. Sumitomo Chemical Company produced alumina fiber by continuously drawing from the solution containing polymerized aluminoxanes [28].

(2) Silica fibers.
Silica fibers were produced by drawing from sols containing $Si(OC_2H_5)_4$ in Asahi Glass Company [29]. The viscous sol becomes drawable when the sol contains long-shaped polymerized fibrous particles. The heat-resistant inorganic fibers are used as supports of high temperature oxidation catalysts.

(3) Silica spheric particles.
Recently, liquid crystal display is very popular. In order to keep the liquid crystal layer at a given thickness (several microns to ten microns), spacers are needed. Originally, high polymer spheres were employed for spacers. Recently, hard spacers were needed for some purposes and silica microspheres which have a diameter larger than several microns are developed in Ube-Nitto Chemical Company [30] and other companies.

ADVANCED MATERIALS FOR FUTURE APPLICATION

Besides the development of commercial products, extensive researches on the sol-gel preparation of advanced materials have been carried out in Japan. Sol-gel processing was applied, whenever a new type of advanced material emerged as key material in high technology. Table IV shows examples of such materials.

(1) Organic photochemical hole burning materials.
Makishima's group applied the sol-gel method to preparation of organic-inorganic microcomposite silica gels containing organic molecules (quinizarin), which show photochemical hole burning at very low temperatures [31]. This work was carried out independently from the work of D.Avnir of Israel, who published the paper on sol-gel preparation of organic-inorganic microcomposites in 1984.

(2) High temperature superconducting oxides.
Soon after the high temperature superconducting oxide was discovered by Bednorz and Müller in 1986, application of sol-gel method to their preparation

Table IV Advanced materials for future application

	Year	Material	Author	Affiliation
(1)	1985	Organic photochemical hole burning	Makishima [31]	University of Tokyo
(2)	1988	High temperature superconducting oxide	Kozuka [32]	Kyoto University
(3)	1988	Ferroelectric (preferentially crystal-oriented)	Hirano [33]	Nagoya University
(4)	1990	NLO: Semiconductor-doped, third order	Nogami [34]	Nagoya Institute of Technology
(5)	1991	NLO: Organic dye-doped, third-order	Nasu [35]	Mie University
(6)	1992	NLO: Gold colloid-doped, third-order	Matsuoka [36]	Mie University
(7)	1993	NLO: Oxide, third-order	Yoko [38]	Kyoto University
(8)	1993	NLO: Oxide, second-order	Hirano [40]	Nagoya University
(9)	1994	Sm^{2+} ion photochemical hole burning	Nogami [41]	Nagoya Institute of Technology
(10)	1995	Solid dye lasers	Yamane [42]	Tokyo Institute of Technology

started in Japan [32]. Fibers and coating films of $YBa_2Cu_3O_{7-x}$ and Bi-(Pb)-Ca-Sr-Cu-O systems were prepared by sol-gel method. The use of sol-gel method successfully lowered the temperature of formation of superconducting oxide phase..

(3) Ferroelectric materials.

Ferroelectric materials have been regarded as importani materials not only for electronics but also for photonic application. Hirano's group prepared $LiNbO_3$ coating films in which crystallites are preferentially oriented [33]. This work triggered preparation of many other ferroelectric and related coating films by Hirano's group and other groups.

(4)—(8) Nonlinear optical materials.

Since late 1980's, material scientists paid much attention to nonlinear optical (NLO) materials for information processing including high speed switching. In Japan, many sol-gel scientists developed different kinds of NLO materials in the form of plate or coating film in 1990's. It was found that sol-gel processing is quite suitable for preparing high performance NLO materials.

Nogami's group prepared third-order NLO discs consisting of silica with CdS [34] and other semiconductor fine particles.

Nasu's group [35] prepared third-order NLO gel doped with organic dye molecules.

Matsuoka et al. [36] prepared third-order NLO silica coating film with dispersed gold colloid, indicating that in sol-gel processing a large amount (e.g. one volume %) of gold can be incorporated. Sakka et al. [37] studied the effect of matrix on the optical properties of gold colloid particles by precipitating them in silica and titania. It was found that absorption peak due to surface plasma resonance shifts to longer wavelengths when the dielectric constant of the matrix is larger. They also prepared silver, platinum and palladium metal colloids in TiO_2 matrix.

Yoko's group [38] prepared third-order NLO oxide films, such as α-Fe_2O_3 [38] and TiO_2 [39]. They showed pretty high third-order NLO susceptibility around 10^{-12} esu. These films are polycrystalline and the distribution of crystals are random, which give the film third-order optical nonlinearity.

Hirano et al. [40] prepared β-BaB_2O_4 coating film in which crystals are preferentially oriented. This film should show a high second-order NLO susceptibility, because the structure of the film is non-centrosymmetric

(9) Sm^{2+} ion photochemical hole burning (PHB).

Sm^{2+} ions in oxides show PHB at room temperature. It is difficult to obtain glasses containing Sm^{2+} ions instead of undesirable Sm^{3+} ions. Nogami et al. [41] showed that sol-gel processing easily gives aluminosilicate glasses in which Sm ions are present in the reduced state, obtaining PHB glasses working at room temperature.

(10) Solid dye lasers.

Usually, dye lasers are solutions and are not so stable to lights. In order to improve the photostability of dye lasers, Yamane et al. [42] prepared organic-inorganic hybrid containing a laser dye DCM. It should be noted that this type of material can be prepared only by sol-gel method.

SUMMARY

An aspect of history of sol-gel technology in Japan was introduced by describing development of commercialized sol-gel materials and advanced future materials. It was shown that many new materials were prepared by sol-gel method, in response to the demands of new technologies.

REFERENCES

1. K.S.Mazdiyasni, R.T.Dolloff and J.S.Smith,II, "Preparation of High-Purity Submicron Barium Titanate Powders," Journal of the American Ceramic Society, 52, 523-526 (1969).

2. R.Roy, "Gel Route to Homogeneous Glass Preparation", Journal of the American ceramic Society,, 52, 344(1969).

3. H.Dislich, "New Routes to Multicomponent Oxide Glasses," Angewandte Chemie, International Edition, 10, 363-370 (1971).

4. S.Sakka, K.Kamiya and I.Yamanaka, "Non-crystalline Solids of the TiO_2-SiO_2 and Al_2O_3-SiO_2 Systems Formed from Alkoxides," pp. 44-48 in Proceedings of the Tenth International Congress on Glass, Kyoto, Volume 13, 1974.
5. "Glasses and Glass Ceramics from Gels," Proceedings of the International Workshop on Glasses and Glass Ceramics from Gels, Padova, Italy, October 8-9, 1981, Journal of Non-Crystalline Solids, 48, 1-230 (1982).
6. K.Susa, I.Matsuyama, S.Sato and T.Suganuma, "New Optical Fiber Fabrication Method," Electronics Letters, 18, 499-500 (1982).
7. S.Sakka, "Fibers from Gels and their Application," p. 114-131 in SPIE Critical Reviews of Optical Science and Technology Vol CR 53, Glass Integrated Optics and Optical Fiber Devices, Edited by S.I.Najafi, 1994.
8. M.Toki, S.Miyashita, T.Takeuchi, S.Kanbe and A.Kochi, "A Large Size Silica Glass Produced by a New Sol-Gel Process," Journal of Non-Crystalline Solids, 100, 479-482 (1988).
9. M.G.M.U.Ismail, Z.Nakai and S.Somiya, "Microstructure and Mechanical Properties of Mullite Prepared by the Sol-Gel Method," Journal of the American Ceramic Society, 70, C7-C8 (1987).
10. T.Hamasaki, K.Eguchi, Y.Koyanagi, A.Matsumoto, T.Utsunomiya and K.Koba, "Preparation and Characterization of Machinable Mica Glass-Ceramics by the Sol-Gel Process," Journal of the American Ceramic Society, 71, 1120-1124 (1988).
11. Y.Shinohara, M.Hirai and M.Ono, "Applications of Ultrathin Alumina Substrates," pp. 481-487 in Advances in Ceramics, Vol 26, The American Ceramic Society, 1989.
12. M.Yamane and M.Inami, "Variable Refractive Index Systems by Sol-Gel Process," Journal of Non-Crystalline Solids, 147 & 148, 606-613 (1992).
13. K.Nakanishi, "Synthesis and Application of Double Pore Silica via Sol-Gel Route," New Ceramics, 9 [8] 68-72 (1996). (in Japanese).
14. H.Yokogawa and M.Yokoyama, "Hydrophobic Silica Aerogels," Journal of Non-Crystalline Solids, 186, 23-29 (1995).
15. S.Ogiwara and K.Kinugawa, "Transparent In_2O_3 Conducting Coating Films Prepared by Thermal Decomposition of Indium Acetyl Acetonate," Journal of Ceramic Society of Japan, 90, 157-163 (1982). (in Japanese).
16. N.Tohge, A.Matsuda, T.Minami, Y.Matsuno, S.Katayama and Y.Ikeda, "Fine-Patterning on Glass Substrates by the Sol-Gel Method," Journal of Non-Crystalline Solids, 100, 501-505 (1988).
17. A.Hattori, K.Makita and S.Okabayashi, " Development of HUD Combiner for Automotive Windshield Application," pp. 272-282 in SPIE Vol 1168, Current Developments in Optical Engineering of Commercial Optics, 1989.
18. M.Murakami, K.Izumi, T.Deguchi, A.Morita, N.Tohge and T.Minami, "SiO_2 Coating on Stainless Steel Sheets from $CH_3Si(OC_2H_5)_3$", Journal of Ceramic Society of Japan, 97, 91-94 (1989). (in Japanese).
19. T.Itoh, H.Matsuda and K.Shimizu, "Black Enhance Color Picture Tube," Toshiba Review, 45 [10] 831-834 (1990). (in Japanese).
20. H.Hayama, T.Aoyama, T.Utsumi, Y.Miura, A.Suzuki and K.Ishiai, "Anti-Reflection and Anti-Static Coating for CRTs," , National Technical Report ,40 [1] 90-96 (1994).

21. S.Yamasaki, H.Inaba, H.Sakai, M.Tatsumisago, N.Tohge and T.Minami, "Water-Repellent Coatings on Glass Substrates by the Sol-Gel Process", pp. 291-295 in Boletin de la Sociedad Espanola de Ceramica y Vidrio, 31-C, Vol 7 (Proceedings of XVI International Congress on Glass), 1992.

22. F.Gunji, T.Yoneda and T.Morimoto, "Novel Functioning of Glass by Wet Coating Process: Part (I) Water-Repellent Glass," New Glass 11 [4] 49-56 (1996). (in Japanese).

23. H.Tomonaga and T.Morimoto, "New Functioning of Glass by Wet Coating Process: Part (II) Ultraviolet Rays Shielding Glass," New Glass 12 [1] 61-67 (1997). (in Japanese).

24. Y.Yoshida, H.Hanaoka, M.Nagata, T.Sakagami and K.Yamada, "Development of Organic-Inorganic Hybrid Coating Agent by Sol-Gel Method", Submitted to Journal of Japan Chemical Society (1988). (in Japanese).

25. T.Yoko, K.Kamiya and S.Sakka, "Preparation of TiO_2 Film by the Sol-Gel Method and its Application to Photoelectrochemical Electrode," Denki Kagaku, 54, 284-285 (1986).

26. Y.Butsugan and K.Niihara, "TiO_2 Coating Photocatalysts with Nano-structure and Preferred Orientation Showing Excellent Activity for Decomposition of Aqueous Acetic Acid," Journal of Material Science Letters, 15, 913-915 (1996).

27. S.Kato, "Development for a Thin Film with Sol-Gel Process," New Ceramics, 9 [8] 28-32 (1996). (in Japanese).

28. S.Horikiri, "Alumina Fibers and heir Application", Ceramics (Japan), 19, 194-200 (1984). (in Japanese).

29. N.Taneda, K.Matsusaki, T.Arai, T.Mukaiyama and M.Ikemura, "Properties and Applications of Silica Fiber by Sol-Gel Process", Reports of Research Laboratory of Asahi Glass Company, 38, 309-318 (1988).

30. K.Toda and T.Adachi, "Silica Spacers," Electronic Materials, June 1996, extra issue, 54-58 (1996). (in Japanese)

31. T.Tani, H.Namikawa, K.Arai and A.Makishima,"Photochemical Hole-Burning Study of 1,4-Dihydroxyanthraquinone Doped in Amorphous Silica Prepared by Alcoholate Method," Journal of Applied Physics, 58, 3559-3562 (1985).

32. T.Umeda, H.Kozuka and S.Sakka, "Fabrication of $YBa_2Cu_3O_{7-x}$ Fibers by the Sol-Gel Method," Advanced Ceramic Materials, 3, 520-522 (1988).

33. S.Hirano and K.Kato, "Preparation of Crystalline $LiNbO_3$ Films with Preferred Orientation by Hydrolysis of Metal Alkoxide," Advanced Ceramic Materials, 3, 503-506 (1988).

34. M.Nogami, K.Nagasaka and E.Kato, "Preparation of Small-Particle-Size, Semiconducting CdS-Doped Silica Glasses by the Sol-Gel Process," Journal of the American Ceramic Society, 73, 2097-2099 (1990).

35. M.Nakamura, H.Nasu and K.Kamiya, "Preparation of Organic Dye-Doped SiO_2 Gels by the Sol-Gel Process and Evaluation of their Optical Non-Linearity," Journal of Non-Crystalline Solids, 135, 1-7 (1991).

36. J.Matsuoka, R.Mizutani, H.Nasu and K.Kamiya, "Preparation of Au-Doped Silica Glass by Sol-Gel Method," Journal of Ceramic Society of Japan, 100, 599-601 (1992).

37. S.Sakka, H.Kozuka and G.Zhao, "Sol-Gel Preparation of Metal Particle / Oxide Nanocomposites," pp. 108-119 in SPIE Vol. 2288 Sol-Gel Optics III, 1994.

38. T.Hashimoto, T.Yoko and S.Sakka, "Third-Order Nonlinear Optical Susceptibility of α-Fe_2O_3 Thin Film Prepared by the Sol-Gel Method," Journal of Ceramic Society of Japan, 101, 64-68 (1993).

39. T.Hashimoto, T.Yoko and S.Sakka, "Sol-Gel Preparation and Third-Order Nonlinear Optical Properties of TiO_2 Thin Films," Bulletin of Chemical Society of Japan, 67, 653-660 (1994).

40. S.Hirano, T.Yogo, K.Kikuta, K.Yamagiwa and K.Niwa,"Processing of β-BaB_2O_4 Powders and Thin Films through Metal Alkoxide," Journal of Non-Crystalline Solids, 178, 293-301 (1994).

41. M.Nogami and Y.Abe, "Sm^{2+}-Doped Silicate Glasses Prepared by a Sol-Gel Process," Applied Physics Letters, 64, 1227-1229 (1994).

42. K.Yagi, S.Shibata, T.Yano, A.Yasumori, M.Yamane and B.Dunn, "Photostability of the Laser Dye DCM in Various Inorganic-Organic Host, Journal of Sol-Gel Science and Technology, 4, 67-73 (1995).

SOL-GEL DERIVED NANOPARTICLES AND PROCESSING ROUTES TO CERAMICS AND COMPOSITES

H. Schmidt, C. Kropf*, T. Schiestel, H. Schirra, S. Sepeur, C. Lesniak[+]
Institut fuer Neue Materialien, Im Stadtwald, Geb. 43 A, D-66 123 Saarbruecken, Germany
*Henkel KGaA, D-40 191 Düsseldorf, Germany
[+]ESK GmbH, D-87 437 Kempten, Germany

ABSTRACT

Highly crystalline nanoparticles from Y-ZrO_2 and FeO_x have been prepared by microemulsion with subsequent solvothermal treatment and by precipitation processes. In both cases, aqueous based intermediates have been prepared which have been surface-modified by carboxylic acids and aminosilane. The surface modification prevents the formation of hard agglomerates of the nanoparticles (6 - 8 nm) completely and provides complete redispersibility. Low sintering compacts and nanocomposites have been prepared, and first results for using the FeO_x nanoparticles for medical applications have been obtained.

INTRODUCTION AND GENERAL ASPECTS OF SOL-GEL NANOPROCESSING

Sol-gel processing has become a very interesting field of research since the first spectacular industrial success has been reported by Schott (Calorex®) based on the investigations of Dislich [1]. The attractiveness of sol-gel processing is based on the fact that inorganic materials can be synthesized by a chemical route having the potential to avoid high processing temperatures such as glass melting (around 1,500 °C or ceramic sintering (from 900 to over 2,000 °C). In general, the sol-gel process is carried out on the basis of soluble precursors such as metal alkoxides or silanes. These precursors are chosen very often since it is possible to obtain homogeneous solution on a molecular level, which then can be reacted to inorganic solids. Whereas in the beginning of the sol-gel technology the majority of papers was devoted to SiO_2, later on ceramic systems also became of interest [2]. Especially based on the work of Hirano [3, 4], of Chen et al. [5] and Payne [6], it could be shown that there is a very interesting potential for low temperature processing. Hirano's and Payne's investigations clearly demonstrated that sintering temperatures can be very low by preforming bonds typical for the later cera-

To the extent authorized under the laws of the United States of America, all copyright interests in this publication are the property of The American Ceramic Society. Any duplication, reproduction, or republication of this publication or any part thereof, without the express written consent of The American Ceramic Society or fee paid to the Copyright Clearance Center, is prohibited.

mic material like in barium titanates, lead titanates, PZT or lithium niobates. This also has been investigated by A. Mosset et al. [7] in the case of barium titanate.

One of the major drawbacks in sol-gel processing of ceramics is related to the low solid content of the green bodies. The low solid content of the green bodies is closely related to the gelation mechanisms. If these coatings are formed, the solid content does not play such an important role because the evaporation of solvents is relatively easy compared to large-size components. However, as pointed out by Fred Lange [8], who investigated the formation of sol-gel films, there might be a critical thickness for obtaining crack-free coatings, and in many cases, this critical thickness is considered to be ≤ 1 µm. Above this thickness, the films, in general, are not able to dissipate the stresses resulting from capillary forces of the drying sol, due to the lack of relaxation ability. This is a result of the strong interactions between inorganic macromolecules (for example silica oligomers) or colloidal particles (e.g. alumina, titania, zirconia and others), leading to highly brittle gels with very low strength. Zarzycki [9] has given experimental data about the strength of silica gel, which is discouragingly weak. In order to improve the processing of gels, several routes have been tried. There are some routes which propose to reduce the interaction of liquids and the pore walls of gels by using solvents other than water or to add components acting as surfactants [10, 11]. Other routes are to reduce the particle-to-particle interaction by surface modification as shown by Schmidt [12]. Surface modification of sol particles may lead to a substantial reduction of the interparticular interaction with the possibility of increasing the solid content of sols and the film thickness, as shown by Mennig [13].

Sols, in general, are stabilized by electric charges by Stern's potential [14]. The "thickness" of the electric charge double layer depends on many parameters, such as ionic strength of the solution, pH value, chemical surface properties of the particles. The transition of an electrostatically stabilized sol into a gel by removal of the electric charges by shifting the system to the point of zero charge or by reducing the sol particle distance below the Stern's repulsion potential, which then changes into attraction, leads to a fast agglomeration reaction, due to the high surface area and the unavoidable presence of reactive groups. This, in most cases, leads to a random distribution of the sol particles which, in general, is far away from a dense packing and, in general, not to a narrow pore size distribution. This may be different in thin films, as shown by Brinker with SAW measurements [15]. The sintering of low density broad pore size distribution gels, in general, leads to a very flat sintering curve with the necessity of using high temperatures for obtaining full density. This can be explained with the fact that areas with small pores sinter at low temperature, even enlarging the large pores, which then have to be densified by high temperatures, and no reasonable advantages are obtained, compared to conventional powder technology. This was one of the reasons that most of the efforts to use sol-gel processing for ceramic parts fabrication have been abandoned.

Recently, nanostructured materials have gained high interest in science as well as in industry. Based on the early fundamental work of Gleiter and his interesting results in ceramics [16, 17], a high industrial potential is attributed to nanostructured inorganic materials. This is a result of the interesting properties of the nanoparticles [12], which is based on their high surface area, quantum size effects, high reactivity, electronic or magnetic size effects or nonlinear-optic effects. Various methods for the fabrication of nanoparticles are investigated. Most of them are related to evaporation condensation methods, plasma methods, laser or other physical methods. A few of them are related to chemical methods like the CVR (Chemical Vapor Reaction) process of H. C. Starck [18], flame spray or flame pyrolysis or hydrolysis processes. In addition to this, wet chemical methods also became of interest, for example, microemulsion processes [19, 20, 21, 22]. One of the drawbacks of the microemulsion techniques, however, is the presence of emulsifiers which, in connection with nanoparticles, in general, contribute substantially to the solid content and, in many cases, are in the range of 50 wt.-%. This causes problems in the processing of the particles or in forming agglomerate-free inorganic powders or in fabricating parts [23]. Other routes are based on the so-called controlled growth process [12, 24] and have shown that in selected systems (zirconia) the particle growth can be controlled by short molecules acting as surfactants [25, 26, 12, 24]. Very good results for the fabrication of agglomerate-free nanopowders have been obtained by the use of oligomers very stably bonded to the surface so that they can even survive hydrothermal treatments, and then, in order to decrease the organic content, to replace them by small molecules [27].

In this paper, several principles for the surface modification of nano-scale particles during or after production are investigated, and it is shown how these particles are able to be used for ceramic and composite materials fabrication.

EXPERIMENTAL

For the synthesis of nanocrystalline Y_2O_3/ZrO_2, microemulsions were prepared by mixing 57.2 ml cyclohexane, 28.6 ml of an aqueous solution of $ZrO(NO_3)_2$ ($C_{(Zr)}$ = 0,4 mol/l) and $Y(NO_3)_3$ ($C_{(Y)}$ = 0,03 mol/l) and 14.3 ml of oleic acid polyethyleneoxide ester ($(CH(CH_2)_7CH{=}CH(CH_2)_7CO_2[(CH_2)_2O]_{20}$, OPE) as emulsifier [20]. After stirring at 34 °C for 3 h stable, transparent microemulsions were obtained. Precipitation was carried out by bubbling NH_3 through the microemulsion until a pH value of 10 was achieved. The water was removed by azeotropic distillation, afterwards the cyclohexane was evaporated at 40 °C and 20 mbar. In order to remove excess OPE, the resulting powder was extracted with 2-propanol in a Soxhlet apparatus. The extracted, modified powder was redispersed in a 1 : 1 water/ethanol mixture (solid content: 10 wt. %) and the resulting suspension was hydrothermally treated (250 °C, 75 bar, 3 h). The solid was separated by centrifugation (4000 r/min, 15 min), after the pH was adjusted to pH = 5.9, the point of zero charge of the powder, and dried at 60 °C and 20 mbar.

In order to exchange the OPE bound on the particles` surface, 50 g of the powders were suspended in 200 ml of 8 N NaOH and 200 ml toluene were added and stirred for 5 h to extract hydrophobic reaction products. The mixture was heated under reflux for 5 h. After the deesterification the solid was separated from the liquid phase after the pH of the aqueous suspension was adjusted to pH = 7 (point of zero charge of ZrO_2) by centrifugation (4000 r/min, 15 min). In order to remove soluble reaction products, the resulting sediment was repeatedly diluted with 1 liter of deionized water, flocculated by pH-adjusting to pH = 7 and filtrated until the conductivity of the washing water was below 5 μS. As surface modifier 4 g of trioxadecanic acid (TODS) were added to the resulting suspension. After stirring for additional 3 h the pH of the suspension was adjusted to pH = 8 (point of zero charge of the TODS-modified ZrO_2) and the flocculated powders were washed as described above in order to remove excess TODS not bound onto the particles` surface. Finally, the powder was isolated by filtration and dried (60 °C, 20 mbar, 5 h). The synthesized powders were characterized by HRTEM, x-Ray, IR-spectroscopy, and laser backscattering. For the fabrication of ZrO_2 containing high refractive index hard coatings, 20 g of the prepared ZrO_2 powders were dispersed in 100 ml distilled water and mixed with a butanolic solution of a boehmite containing condensate with epoxysilane (hard coating basis material) as described in detail in [28, 29] up to a concentration of 30 - 40 wt.-% of ZrO_2. The clear liquid was used to coat polycarbonate lenses pretreated by a plasma in a dip coating procedure. After drying, the coating was cured at 130 °C for 2 hrs.

For the preparation of nano-scale iron oxide, 89.40 g of ferrous chloride tetrahydrate ($FeCl_2*4H_2O$) and 243,3 g of ferric chloride hexahydrate ($FeCl_3*6H_2O$) were dissolved in 2 liters of oxygene-free deionized water. 5 M NaOH was added until a pH of 11.5 is reached. The mixture was heated up to 65 °C for 10 min. The resulting black precipitate was washed repeatedly with deionized water. For surface modification, to an acidic (pH 5) aqueous suspension of iron oxide (5 wt.-%) γ-aminopropyl triethoxysilane (APS) was added until a weight ratio silane to iron oxide of 0.8 was reached. The suspension was pourred into ethyleneglycol (water/ethyleneglycol; v/v=1), heated up to 80 °C and sonified for 48 hours. Subsequently, water and ethanol were distilled off under vacuum at 50 °C. The glycolic suspension was centrifuged for 60 min at 2500 m/s², and the supernatant colloidal suspension was dialyzed against deionized water. The as prepared particles were characterized with respect to particle/agglomerate size by laser light scattering and TEM. Surface chemical properties were determined by zeta potential measurements and their composition were analyzed by ICPAES.

RESULTS

Zirconia and Zirconia Composites

As shown elsewhere [30], microemulsion systems based on ZrO_2 and yttria-ZrO_2 have been used for the fabrication of zirconia nanoparticles. However, the removal of the emulsifiers by oxidation at elevated temperatures leads to strongly agglomerated powders. For this reason, a method has to be developed, which allows the preparation of well crystallized powders at temperatures where no agglomeration takes place. Solvothermal treatment in the presence of surface protecting components seemed to be an adequate means. A microemulsion has been prepared from zirconium and yttrium nitrates in cyclohexane by dispersing the aqueous nitrate solution in the organic solvent in the presence of OPE. OPE was chosen since it was known to be an effective emulsifier as well as to adsorb strongly to oxide surfaces. After three hours stirring, completely transparent microemulsions have been obtained. At this temperature, relatively high water-to-cyclohexane ratios could be realized, as described in [31]. The emulsifier concentration, however, is rather high and requires about 15 % by volume of the total solution, which leads to a rather high organic content in the whole system. The droplet size under these conditions is 13 nm. Precipitation within the droplets is obtained by NH_3 bubbling. After the removal of cyclohexane and water by vacuum treatment and azeotropic distillation, the particle size is in the range between 2 and 3 nm, as shown in fig. 1 from the TEM micrograph.

The X-ray analysis of these powders clearly indicates that the crystallinity of these precipitates is rather poor. Due to the poor crystallinity, a solvothermal treatment at 250 °C for three hours was carried out, and after the treatment, the separation of the powder from the solvent was performed by repeatedly washing and centrifugation, leading to NH_4NO_3-free powders. After centrifugation the FTIR spectra were taken (fig. 2). They clearly show that the OPE still is linked to the surface of the particle and the photon correlation spectroscopy for the determination of the particle size shows that the particle size has increased to about 3 - 10 nm, indicating a growth reaction in the presence of OPE (figs. 3 and 4) [31]. The powders can be redispersed in water completely agglomerate-free.

20 nm

Fig. 1: TEM micrograph of the OPE coated Y-ZrO_2 powder after the microemulsion preparation.

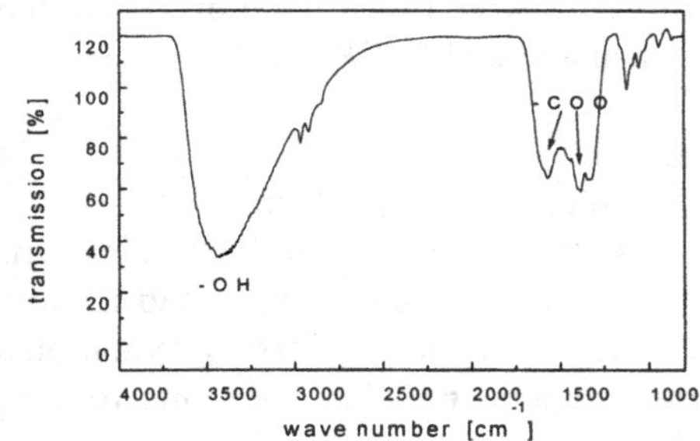

Fig. 2: IR-spectrum of Y_2O_3/ZrO_2 nanoparticles coated with OPE after precipitation from the microemulsion..

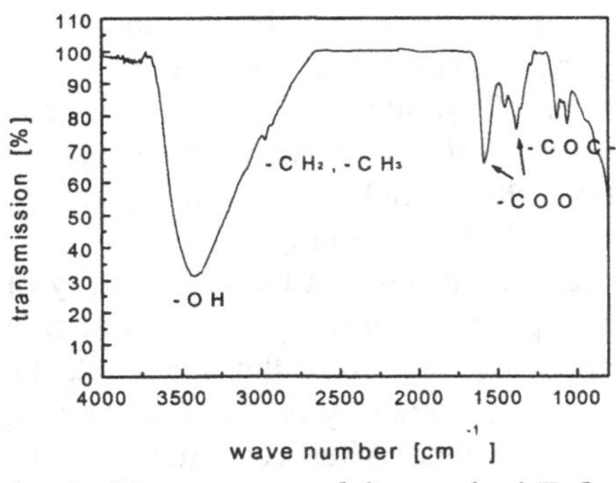

Fig. 3: IR-spectrum of the washed ZrO_2 nanoparticles coated with OPE after hydrothemal crystallization at 250 °C, 75 bar, 3 h.

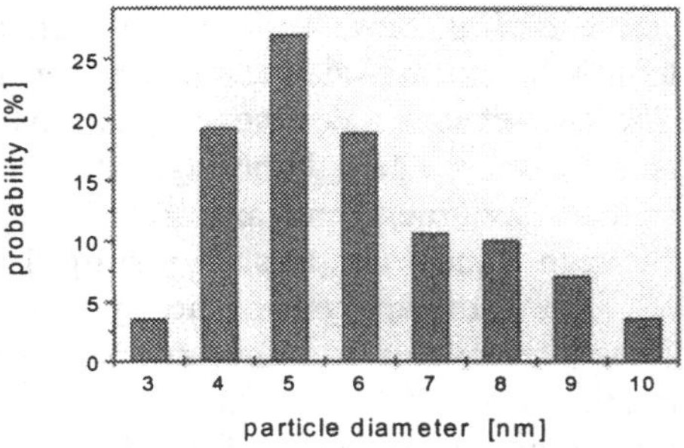

Fig. 4: Particle size distribution calculated from HRTEM micrograph of hydrothermal crystallized Y_2O_3/ZrO_2 (250 °C, 75 bar, 3 h) surface modifier coated with OPE.

In fig. 5 the HRTEM micrograph of these powders is shown, and fig. 6 shows that these powders now exhibit a good crystallinity (cubic modification), the line broadening of which matches with the observed particle size. This indicates that mass transport as well as crystallization can take place under solvothermal conditions under the surface protecting effect of OPE.

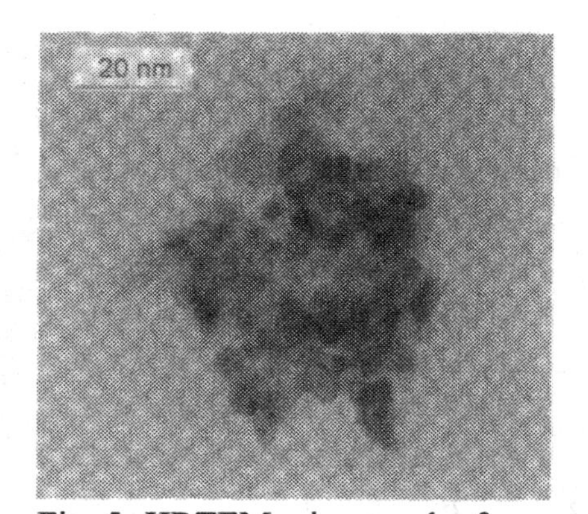

Fig. 5: HRTEM micrograph of Y_2O_3/ZrO_2 after hydrothermal crystallization (250 °C, 75 bar, 3 h), surface modifier: OPE.

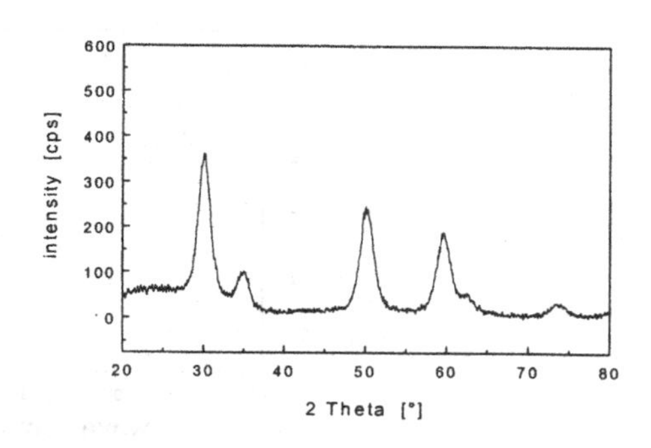

Fig. 6: X-ray diffraction spectrum of Y_2O_3/ZrO_2 after hydrothermal crystallization (250 °C, 75 bar, 3 h).

In [27] it was shown that Y-ZrO_2 powders obtained from controlled growth precipitation processes covered with polymeric protective components could be reacted with smaller molecules to reduce the organic content without losing their surface protection. For this reason, a process was employed, the experimental of which was described elsewhere [27]. The OPE-modified particles were treated with a two-phasic mixture of toluene and sodium hydroxide. The sodium hydroxide removes the OPE, which then is subsequently dissolved in the toluene phase, and at the same time, a negative ζ-potential is established on the zirconia particle surface in order to avoid aggregation. After changing the pH to 7, flocculation takes place in the system, and it can easily be separated by centrifugation from the liquid. The powder repeatedly was washed with deionized water. After washing to neutral pH, carboxylic acid (trioxadecanic acid) was added, which led to a complete redispersion of the powder, and as investigated by photon correlation spectroscopy, the particle size distribution remains unchanged compared to the state after the hydrothermal treatment. These powders are completely redispersible in water as well as in ethanol, and it could be shown by DRIFT spectroscopy that the carboxylic acid is bonded to the particle surface (fig. 7), since the position of the =C=O peak represents the carboxylate ion. The process shows that the flocculated system is only weakly agglomerated and can be recoated by carboxylic acids before hard agglomerates are formed. The total organic content now is reduced to < 5 wt.-%, representing an almost 75 % coverage of the nanoparticle surface. As also shown in [27], the ζ-potential is shifted to positive values.

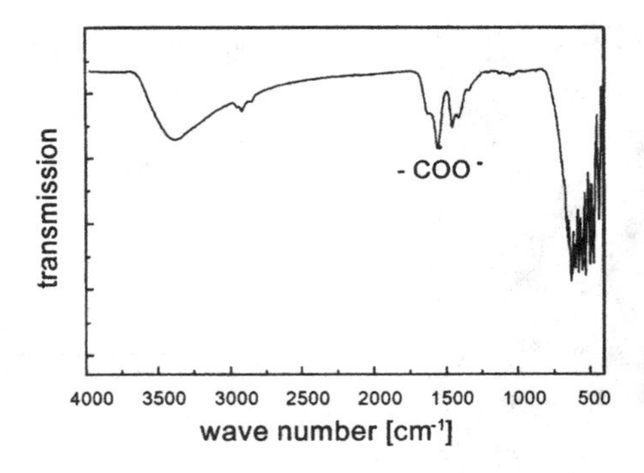

Fig. 7: IR-spectrum of the TODS-modified Y_2O_3/ZrO_2.

Compacts prepared from these powders by cold isostatic pressing can be sintered to full density at 1130 °C, as shown in fig. 8. The pores of the green bodies made from these powders measured by BET show a narrow size distribution with a maximat 3.5 nm, which is in a good agreement with the values calculated from the particle size distribution. The sintering curve of the green body (fig. 8b) shows a one step sintering behavior which was finished at a temperature of 1100 °C.

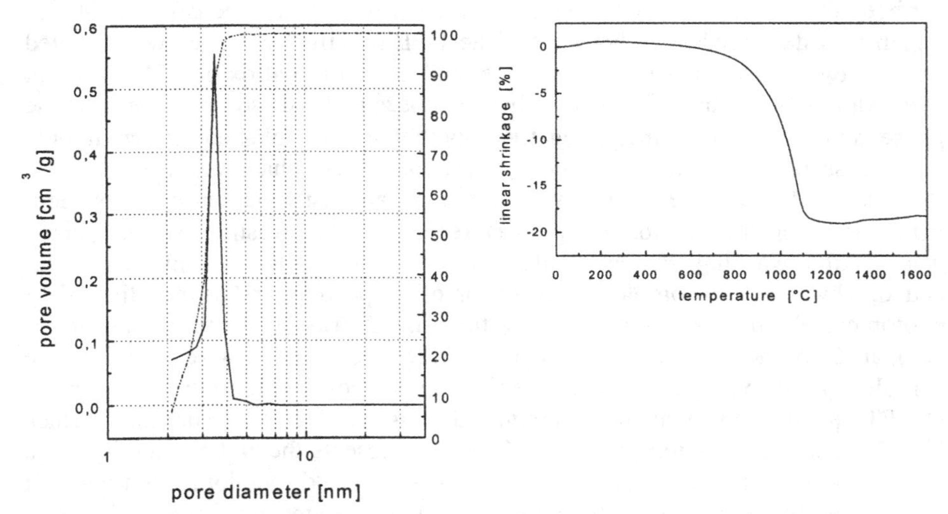

Fig. 8a: Pore size distribution of cold pressed hydrothermally crystallized Y_2O_3/ZrO_2 powder.

Fig. 8b: Sintering curve of cold pressed hydrothermally crystallized Y_2O_3/ZrO_2 powder (96 % density).

The low sintering temperature obtained from cold pressing which is, in general, not the most favorable way to produce green bodies from nanopowders, is attributed to the effect of ODS as pressing aid leading to a monomodal nanoporosity. The yield of the microemulsion/solvothermal process sums up to 3.4 g powder per 100 ml of reaction volume. Due to the solvothermal treatment, no high temperature calcination step is necessary to obtain highly crystalline powders. The obtained results show that OPE can be used as an emulsifier and, at the same time, as protecting agent during the hydrothermal process. Through the washing steps, all soluble salts such as nitrates can be removed completely. Depending on the type of processing, solid contents of green compacts up to 50 % by volume could be fabricated. The highest green densities could be obtained by a compounding process with subsequent extrusion [32].

For the fabrication of transparent hard coatings, these powders were suspended in water and formed a clear solution. This solution was mixed with a hard coating system described in [28]. Coatings on polycarbonate eye glass lenses have been fabricated by dip coating of the plasma pretreated lenses. Plasma treatment leads to a good adhesion on the substrates surface with cross-cut/tape test according to DIN 53151-A-B and 58196-K2 show values of CC/TT = 0-1/0-1. The abrasion resistance of the zircon dioxide-modified hard coatings (film thickness: 3 - 5 μm) was determined by the Taber Abraser test according to DIN 52347 (1,000 cycles/abrasion wheels: CS-10F/5,4 N). The wear resistance of a system containing 30 % zircon dioxide is down to ≈ 5 % haze after 1,000 cycles. The refractive index is close to 1.6.

In order to fabricate iron oxides, which basically can also be carried out by micro-emulsion techniques [33]), a more simple route was investigated. There is a large body of literature [34, 35, 36, 37, 38] available which shows that iron oxide nanoparticles can be easily precipitated from aqueous solutions. However, the problem of agglomeration is still difficult according to the state of the art. Since it can be assumed that in the presence of water the agglomeration of iron oxide particles is weak, a step was anticipated to deagglomerate the iron oxide in presence of surface modifiers. In fig. 9 and fig. 10 the particles and the particle size distribution of iron oxide as precipitated is shown. The primary particle size ranges from 8 - 12 nm, however, the particle size obtained from laser back scattering ranges from 30 - 100 nm which is due to agglomerates still stable under measuring conditions.

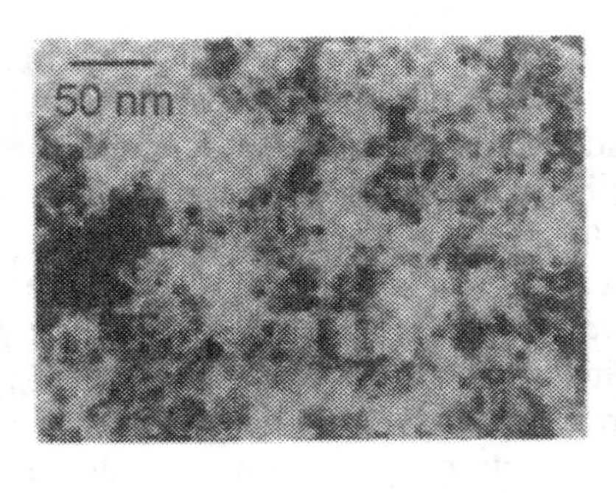

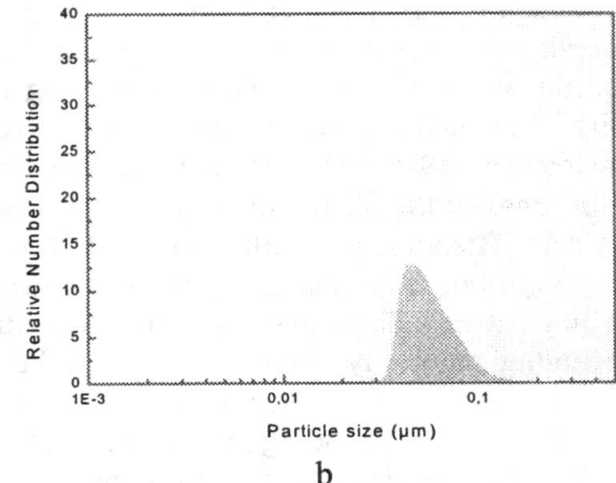

a

b

Fig. 9: TEM micrograph of unmodified iron oxide particles.

Fig. 10: Particle size distribution of iron oxide in aqueous suspension (determined by laser light scattering).

According to x-ray analysis (fig. 11), the particles consist of either magnetite or maghemite. Both nanocrystalline phases cannot be distinguished due to the almost identical lattice parameters. From the peak broadening, a crystallite size of 9 nm was calculated using the Scherrer equation. Due to the small particle size, superparamagnetic behaviour is expected. Measurements in a vibrating sample magnetometer revealed a saturation magnetization of 68 EMU/g (bulk magnetite 122 EMU/g, bulk maghemite 108 EMU/g). The absence of a hysterisis loop indicates superparamagnetic properties (fig 12). For deagglomeration and stabilization of the FeO_x particles, the treatment of the slurry with γ-aminopropyl triethoxysilane was carried out as described in the experimental.

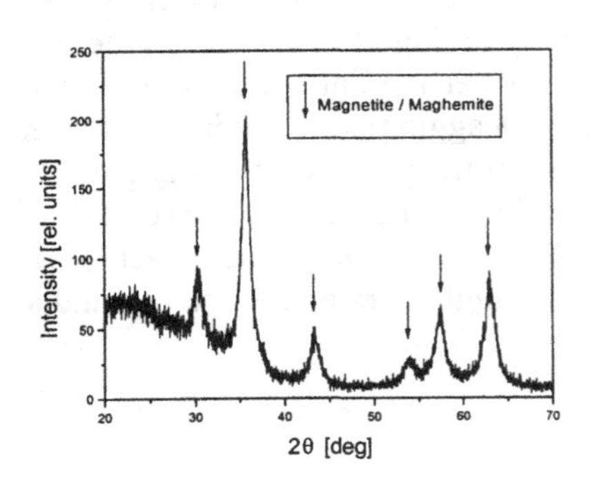

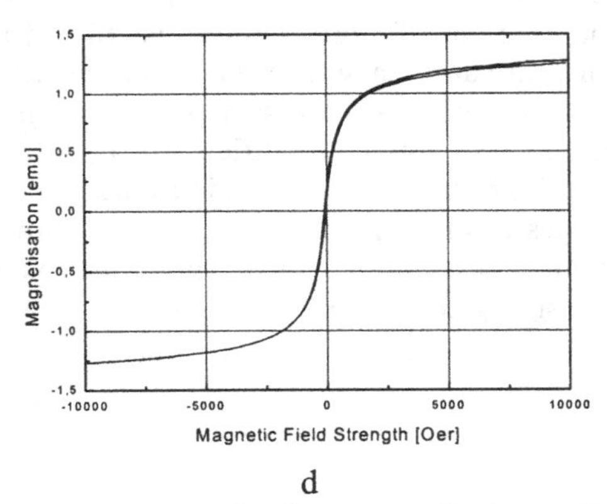

c

d

Fig. 11: X-ray diffraction diagramm of iron oxide powder.

Fig. 12: Magnetization curve for unmodified iron oxide particles.

In fig. 13 the particle size distribution determined with laser backscattering is shown before and after surface modification. The average particle size is shifted from 50 nm to 10 nm. Compared with the primary particles size derived from TEM micrograph (see fig. 9) it can be concluded that the particles are completely deagglomerated. In additon no agglomerates can be found on TEM micrographs after surface modification (fig. 14).

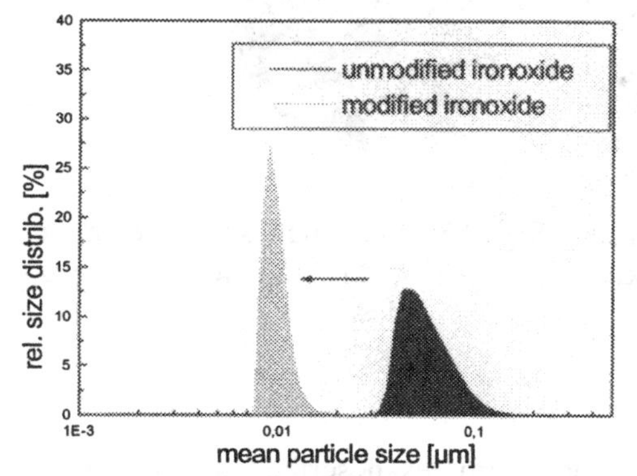

Fig. 13: Comparison of particle size distribution for unmodified and modified iron oxide (measured by laser light scattering).

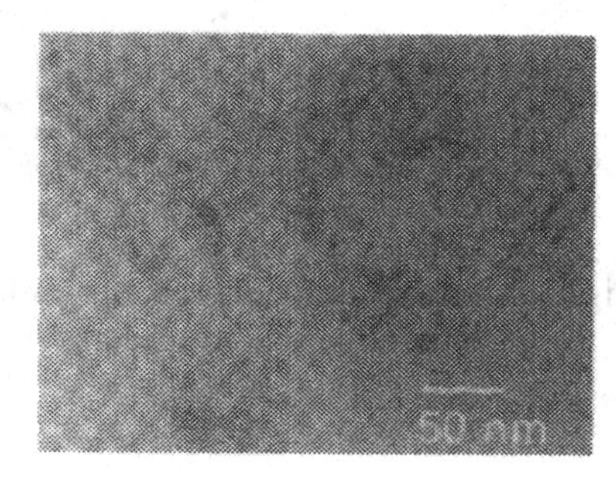

Fig. 14: TEM micrograph of a silane modified iron oxide.

To demonstrate the effect of surface modification on the surface chemical properties, ζ-potential measurements of unmodified and modified iron oxide were compared (fig. 15). After the modification, the particles possess an isoelectric point of 9.5 (unmodified particles IEP of 6), indicating the presence of free amino groups on the particle surface. A silane content between 4 and 5 rel. wt.-% was determined by chemical analysis. From this value a monomolecular layer around the iron oxide particle can be calculated (average molecule size aminosilane appr. 60 A^2). Fig. 16 shows a structure model for the prepared particles. Aqueous suspensions of functionalized iron oxide are stable against agglomeration for a long time (> 6 months) and a desorption of the coating was not observed.

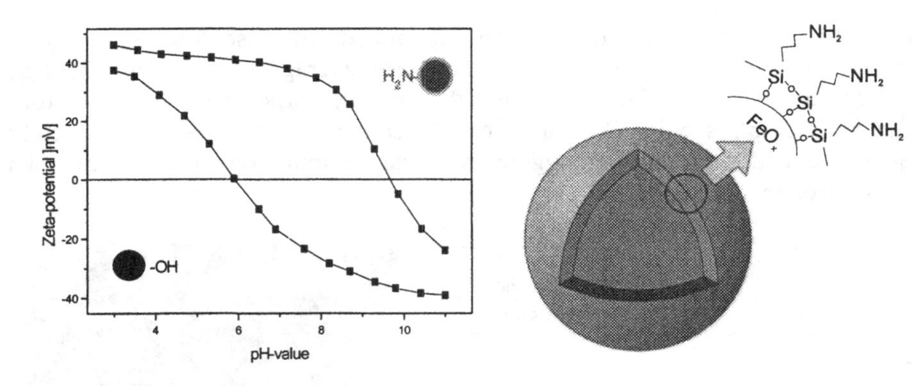

Fig. 15: Zetapotential vs. pH curves for unmodified and modified iron oxide.

Fig. 16: Core-shell structure model for the functionalized particles.

Medical Applications

The superparamagnetic iron oxide nanoparticles described above possess interesting properties for medical applications. Their size in combination with suitable surface chemical properties makes a recognition by reticuloendothelial system difficult. This allows an application as drug carrier systems. Their magnetic properties can be used for a new hyperthermic cancer therapy, which is based on the following principles:

Superparamagnetic particles get magnetized in a magnetic field. If the field is switched off the particles dissipate their magnetic energy into thermal energy. If these particles are brought into tumor tissue a controlled heating is possible by applying an external AC magnetic field. This therapy is called "magnetic fluid hyperthermia".

In in-vitro experiments aminosilane modified iron oxide nanoparticles the details of which are described elsewhere [39, 40] showed a higher uptake into malignant cells (fig. 17) compared to normal cells. The tumor cells eagerly consume these nanoparticles but are not able to digest them. Irradiation by an alternating magnetic field led to heating up of the cells and their destruction.

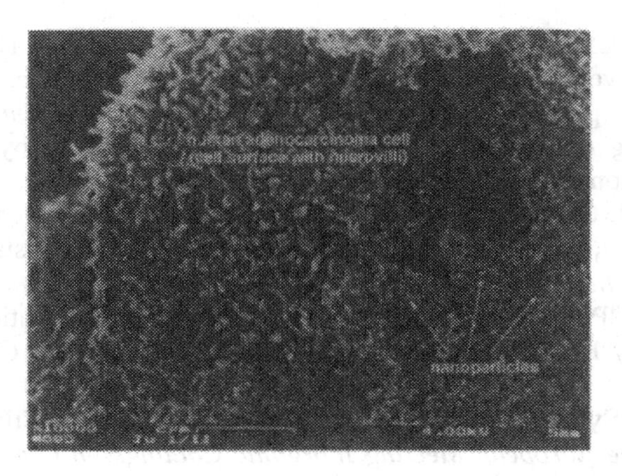

Fig. 17: REM picture of a malignant cell in the presence of aminosilane modified iron oxide particles.

CONCLUSIONS

The experiments show that sol-gel preparation techniques carried out under "protective“ conditions can be used to prepare agglomerate-free nanoparticles with specific surface functions. Using this approach, nanoparticles can be used advantageously for ceramic parts processing, for polymer matrix nanocomposite hard coating fabrication and for medical application.

ACKNOWLEDGEMENT

The authors want to thank Mr. D. Burgard and Mr. N. Bendzko for their helpful discussions, the European Commission Directorate-General XII, the Ministry for Research and Culture of Saarland and the Federal Ministry for Research and Technology for their financial help.

REFERENCES

[1]H. Dislich, Angew. Chem. 83, p. 428 (1971).

[2]C. J. Brinker, and G. W. Scherer, Sol-Gel Science, Academic Press, London, 1990.

[3]S. Ono and S.-I. Hirano, "Synthesis of Highly Oriented Lithium Niobate Thin Film from Neutralized Aqueous Precursor Solution“, *J. Am. Ceram. Soc.*, **80** [11] 2869-2875 (1997).

[4]S.-I. Hirano and K. Kato, "*Manufacture of Lithium Niobate Powder*“, Japanese Patent No. JP 01 09 34 25 A2, April 12, 1989.

[5]E. Wu, K.C. Chen, J.D. Mackenzie, „Ferroelectric Ceramics - The Sol-Gel Method versus Conventional Processing“, *Mat. Res. Soc. Symp. Proc.* **32**, 169 (1984).
[6]D.A. Payne, p. 39 in *Proceedings International Symposium on Molecular Level Designing of Ceramics*, Nagoya, March 1991. Edited by Team on the NEDO International Joint Research Project, 1991.
[7]A. Mosset, I. Gautier-Luneau, J. Galy, P. Strehlow and H. Schmidt, "Sol-Gel Processed $BaTiO_3$ - Structural Evolution from the Gel to the Crystalline Powder“, *J. Non-Cryst. Solids* **100**, 339 - 344 (1988).
[8]F.F. Lange, "Processing-Related Fracture Origins: I, Observations in Sintered and Isostatically Hot-Pressed Al_2O_3/ZrO_2 Composites“, *J. Am. Ceram. Soc.* **66**, 396-398 (1993)
[9]J. Zarzycki, Synthesis of Glasses from Precursor: Bulk and Film - a Comparison, *Proc. of the European Meeting Inorganic Coatings on Glass.* Edited by P. Picozzi, S. Santucci, P. Boattini, L. Massarelli and V. Scopa, Soietà Italiana Vetro, L'Aquila, Italy, 149 (1988).
[10]R.K. Iler, „The Chemistry of Silica“. J. Wiley & Sons, New York, 1979.
[11]S. Wallace and L.L. Hench, "The Processing and Characterization of CDDA Modified Gel-Derived Silica“, in: *Better Ceramics Through Chemistry*. Edited by C.J. Brinker, D.E. Clark, D.R. Ulrich. North-Holland, New York, 19984.
[12]H. Schmidt, KONA Powder and Particle, No. **14**, 92 - 103 (1996).
[13]M. Mennig, G. Jonschker and H. Schmidt, "Sol-Gel Derived Thick SiO_2 Coatings and Their Thermomechanical and Optical Properties, *SPIE Proc. Sol-Gel Optics* **1758**, 125 - 134 (1992).
[14]O. Stern, *Z. Elektrochem.*, **508** (1924).
[15]G.C. Frye, A.J. Ricco, S.J. Martin and C.J. Brinker, "Characterization of the Surface Area and Porosity of Sol-Gel Films Using SAW Devices“, pp. 349 - 354 in *Better Ceramics Through Chemistry*. Edited by C.J. Brinker, D.E. Clark and D.R.Ulrich, Materials Research Society, Pittsburgh, PA, 1988.
[16]H. Gleiter, Nanocrystalline Materials, Pergamon Press, Oxford, 1989.
[17]H. Gleiter, R. Birringer and J. Karch, "Verfahren zum Herstellen eines plastisch verformbaren keramischen oder pulvermetallurgischen Werkstoffes und unter Anwendung eines solchen Verfahrens hergestellter Gegenstand“, European Patent No. 0 317 945 B1, Nov. 22, 1988.
[18]R. L. Meisel, T. König, in: Werkstoffwoche 1996, Symp. 9 Neue Werkstoffkonzepte, edited by H. Schmidt and R. F. Singer, (DGM Informationsgesellschaft mbH, Frankfurt/M., 1997).
[19]K. Osseo-Asare and F. J. Arriagada, Ceramic Transactions 12, Ceramic Powder Science III, edited by G. L. Messing, S. Hirano and H. Hausner, (American Ceramic Society, Westerville / Ohio, USA, 1990) 3-16.

[20]S. D. Ramamurthi, Z. Xu und D. A. Payne, *J. Am. Ceram. Soc.* **73**, 2760-63 (1990).
[21]R. Naß, D. Burgard, H. Schmidt, in: "Proceedings of the 2nd European conference on Sol-Gel Technology", edited by R. Naß, H. Schmidt and S. Vilminot, North Holland Publishers, Amsterdam, The Netherlands (1992).
[22]D. Burgard, Master Thesis, University of Saarbrücken, Germany 1992.
[23]H. Herrig. R. Hempelmann, *Mater. Lett.*, **27** [6], (1996), 287-292.
[24]D. Burgard, R. Naß and H. Schmidt, Synthesis and colloidal processing of nanocrystalline (Y_2O_3 stabilized) ZrO_2 powders by a surface free energy controlled process, in *Mat. Res. Soc., Symp. Proc.,* Pittsburgh/PA, 432:113 (1997).
[25]E. Scolan, J. Maquet, C. Bonhomme, F. Ribot and C. Sanchez, "Synthesis-Characterization-Reactivity of Titanium Oxo Nano-Building Blocks: from Oxo-Alkoxo Titanium Clusters to TiO_2 Nanoparticles", Proc. MRS Spring Meetings, April 13 - 17, 1998, San Francisco (in print).
[26]R. Naß and H. Schmidt, Formation and properties of chelated alumium-alkoxides, in: *Ceramic Powder Processing Sciencs.* Edited by H. Hausner, G. L. Messing and S. Hirano, Deutsche Keramische Gesellschaft e. V., Köln, 69 - 76 (1989).
[27]H. Schmidt, R. Nass, D. Burgard and R. Nonninger, "Fabrication of Agglomerate-Free Nanopowders by Hydrothermal Chemical Processing", Proc. MRS Spring Meeting, April 13 - 17, 1998, San Francisco (in print).
[28]R. Kasemann, H. Schmidt and E. Wintrich, "New Type of a Sol-Gel-Derived Inorganic-Organic Nanocomposite", in *Mat. Res. Soc. Symp. Proc.* **346**, 915 - 921 (1994).
[29]H. Schmidt, E. Arpac, H. Schirra, S. Sepeur and G. Jonschker, "Aqueous Sol-Gel Derived Nanocomposite Coating Materials", Proc. MRS Spring Meeting, April 13 - 17, 1998, San Francisco (in print).
[30]D. Burgard, Master's Thesis, University of Saarland, Saarbruecken, Germany (1991).
[31]C. Kropf, Ph. D. Thesis, University of Saarland, Saarbruecken, Germany (1998).
[32]to be published later.
[33]to be published later.
[34]R. Massart, "Preparation of Aqueous Magnetic Liquids in Alkaline and Acidic Media", *IEEE Trans. Magn.* **17**[2], 1247-1248 (1981).
[35]H. Pilgrimm, "Stabile magnetische Flüssigkeitszusammensetzungen und Verfahren zu ihrer Herstellung und ihre Verwendung", German Patent DE 3709852, 1984.

[36]R.E. Rosensweig, "Ferrohydrodynamics", Cambridge University Press, Cambridge, 1985.
[37]T. Sato, T. Iijima, M. Seki, N. Inagaki, „Magnetic Properties of Ultrafine Ferrite Particles", *J. Magn. Magn. Mat.* **65,** 252-256 (1987).
[38]N.M. Gribanov, E.E. Bibik, O.V. Buzunov, V.N. Naumov, "Physico-Chemical Regularities of Containing Highly Dispersed Magnetite by the Method of Chemical Condensation", *J. Magn. Magn. Mat.* **85,** 7-10 (1990).
[39]A. Jordan, P. Wust, R. Scholz, H. Faehling, J. Krause, R. Felix, "Magnetic Fluid Hyperthermia", in: *Scientific and Clinical Applications of Magnetic Carriers.* Edited by Häfeli et al., Plenum Press, New York, 1997, 569 - 595.
[40]A. Jordan, R. Scholz, P. Wust, H. Faehling, J. Krause, W. Wlodarczyk, B. Sander, T. Vogl, R. Felix, "Effects of Magnetic Fluid Hyperthermia (MFH) on C3H Mammary Carcinoma *in vivo*", *Int. J. Hyperthermia*, Vol. **13**, [6], 587 - 605 (1997).

SOL-GEL PROCESS FOR OPTICAL FIBER MANUFACTURE

John MacChesney, D. W. Johnson, Jr., S. Bhandarkar, M. Bohrer,
J. W. Fleming, E. M. Monberg, and D. J. Trevor
Bell Laboratories
700 Mountain Avenue
Murray Hill, NJ 07974

ABSTRACT

Sol-gel has been pursued as a means for producing vitreous silica of high intrinsic worth for optical fiber manufacture. These efforts took various routes, each seeking net-shape forming at room temperature to make bodies meeting exacting performance requirements. Bell Laboratories developed a gel-casting process to produce large tubes to serve as jacketing which comprises more than 90% of the fiber's mass. Worldwide fiber production consumed greater than 1000 metric tons of fused silica in 1997, suggesting a significant market for inexpensive waveguide-quality products.

INTRODUCTION

During the past decade approximately ten thousand papers have been devoted to sol-gel processing. Yet it is one of the oldest processing means. Clay sols were used to produce ancient pottery. In the late 1940's, Rustum Roy, et al. [1] used oxide sols to produce ultra-homogeneous samples for phase-equilibria studies. Later almost frenetic activity occurred in both industrial and academic laboratories throughout the world to develop sol-gel as a processing means for films, coatings, abrasive grain and monolithic bodies. These efforts tended to target high-end markets. One area, in particular, is that of vitreous silica for optical fibers. The objective was to lower manufacturing costs of silica glass having precise shape, ultimate purity, and high mechanical strength which is required for fibers.

These efforts included both the hydrolysis-polycondensation of silicon alkoxides and the gelation of colloidal sols. The latter has proven the most efficacious primarily because it allowed higher loading of silica in the sol resulting in a strong green body capable of withstanding drying stress. The colloidal approach employed many of the conventional ceramic forming processes: isostatic pressing [2], extrusion [3], electrophoretic deposition [4], centrifugation [5], and gel casting [6]. The latter has been pursued by Lucent Technologies towards commercial production, with one goal being to produce

To the extent authorized under the laws of the United States of America, all copyright interests in this publication are the property of The American Ceramic Society. Any duplication, reproduction, or republication of this publication or any part thereof, without the express written consent of The American Ceramic Society or fee paid to the Copyright Clearance Center, is prohibited.

large silica tubes, nominally 50 mm in diameter and one meter long to overclad MCVD preforms. In this configuration the fiber's optical path would remain in the vapor-deposited silica provided by MCVD, but approximately 90% of the mass of the eventual fiber is provided by the gel tube.

Processing is described in U.S. Patent 5,379,364 [7] and outlined in Figure 1. Here colloidal silica particles made by flame-hydrolysis of silicon chlorosilanes (OX-50, Degussa AG, Frankfurt, Germany) are dispersed in water, stabilized at a high pH and centrifuged to remove impurity particles. Next, an ester is added to the sol and it is cast in a mold to gel. After an appropriate time for equilibration, the gel is pushed from the mold in water, positioned on rollers, removed and rotated slowly in a controlled atmosphere to dry. The porous silica bodies are next purified by heating to (1) remove organics and residual water, (2) to dehydrate the body in a chlorine containing atmosphere, (3) treatment in thionyl chloride to remove refractory oxides. The purified porous cylinder is then sintered to pore-free vitreous silica by pulling it through a furnace heated to 1500°C. Finally, it is fused over an appropriate MCVD core-rod.

- Dispersion of fumed silica in water
- Stabilization of sol with organic base
- Centrifugation to remove particulate impurities
- Deaeration
- Addition of ester
- Casting of gel body in precision mold
- Remove mandrel and launch tube
- Drying
- Purification
- Sintering
- Characterization

Figure 1. Processing steps for sol-gel preparation of silica tubes.

This design resulted in a fiber where the optical power in the core was reduced by several orders of magnitude at the core-rod/overclad interface. Therefore, transition metal ions in the overcladding tube would have a minimal effect on optical loss. In spite of this precaution, Neutron Activation Analysis of the gel-silica shows that concentrations of Fe, Co, Ni, etc. are below 100 ppb. In fact, spectral loss measurements at 1310 nm on experimental fibers recorded loss between 0.39 and 0.33.

Design could not overcome the degradation of fiber strength by the existence of internal flaws caused by refractory particles [8] such as zirconia or chromia which occur in the starting material or are added during processing. These particles, as small as 1 nm in diameter can cause fiber to break at 100 Ksi proof test levels. To meet current commercial standards these must be reduced to less than one break per 1000 Km of fiber or about one part in 10^{16}. To accomplish this, sol processing and casting are carried out in class 1000 clean rooms. Centrifugation of the sol removes the majority of particles. Finally, $SOCl_2$ [9] preferentially attacks these oxides in the porous silicon matrix during the purification step.

However, the achievement of optical fiber quality silica is only half the story. To make fiber meeting current commercial standards, precise net-shape formed cylinders are required. Ultimately, these determine the dimensions of the fiber made from them as well as certain optical properties. Tube dimensions depend upon mold finishing and care in the processing of the friable porous silica bodies formed in them. Finished sintered tubes are normally 115 cm long, the 154 cm dried gel body having shrunk 25% during sintering. Other dimensional characteristics of 185 experimental tubes, 1 meter long made for testing are as follows:

OD variation	0.6%
ID variation	0.7%
Ave. Concentricity/OD	$0.15 \pm 0.03\%$
Max Ovality/OD	$0.08 \pm 0.06\%$
Straightness	0.59 ± 0.33 mm/m

where: concentricity is the displacement of the ID center from the center of the OD; ovality is the difference between the maximum and minimum OD; and straightness is the maximum deviation of the outside surface from an ideal straight cylinder.

DISCUSSION

Successful fabrication of large monolithic structures depends primarily upon achieving a gel body with sufficient strength to withstand stresses created during drying as well as those encountered in processing. Alkoxide derived gels are poorly suited to this task because the loading (wt % silica in the sol) is quite low, below 20% by most recipes. Drying of these gels results in significant shrinkage and drying stress. Colloidal gels made with relatively large particles like those comprising OX-50, can be loaded to about 50% silica. These bodies are stronger, shrink less and are better candidates for surviving processing.

Thus, for the 50 m^2/g silica used in this study, the sol was prepared by mixing fumed silica with water and adding an organic base (such as tetramethyl ammonium hydroxide) to raise the pH well into the sol stability range for amorphous silica shown in Figure 2 taken from Iler[(10)]. At a pH of 10-12, the sol is very stable and can be stored for months without flocculation. Gelation occurs when the sol is mixed with a hydrolyzable ester such as methyl formate and cast into a mold. The methyl formate reacts with water according to the reaction:

$$HCOOCH_3 + H_2O \Rightarrow HCOOH + CH_3OH$$

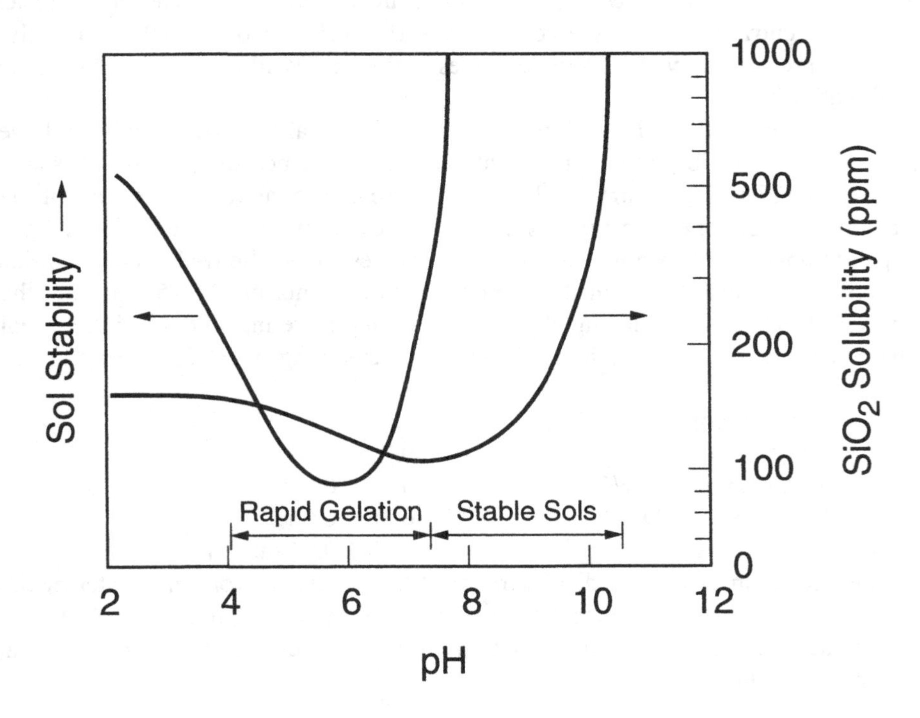

Figure 2. Amorphous silica sol stability and solubility as a function of pH. Adapted from Iler[10] Figures 1.6 and 4.13.

The reaction product, formic acid, serves to reduce the pH of the sol and gelation. The advantage of using a hydrolyzable ester rather than adding an acid directly comes from the slow hydrolysis which allows the sol to be mixed, poured into a mold and deaired before gelling. Direct addition of an acid would result in immediate gelation in the vicinity of the acid and prevent the casting of a fluid, homogeneous sol. The reduction in pH of a sol with time is shown in Figure 3 for a sol stabilized with tetramethylammonium hydroxide [$(CH_3)_4NOH$] to which methyl formate was added in the ratio of 1.1 moles of methyl formate for each mole of tetramethylammonium hydroxide. As can be seen in Figure 3, the sol gels in 5-6 minutes (at room temperature) at pH 8.8. The pH continues to drop and can reach values between 6-7 after long times. At lower pH, syneresis occur, the cylinders shrink, and are easily removed from the molds because they are rigid enough to keep their shape and yet plastic enough to resist brittle fracture.

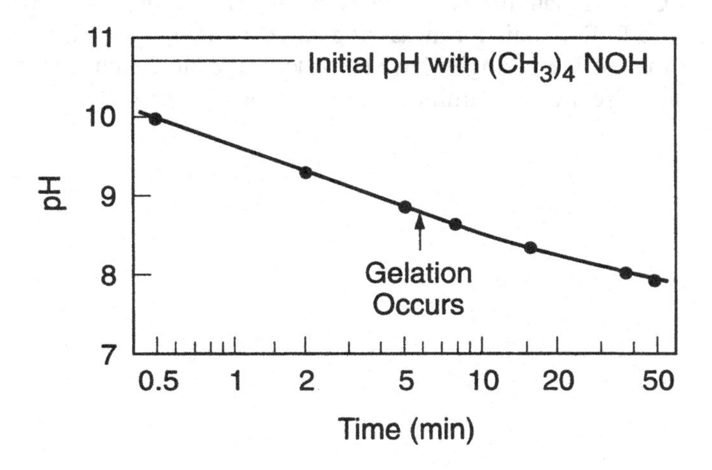

Figure 3. Reduction of pH with time (at room temperature) after addition of methyl formate to a sol stabilized with tetramethylammonium hydroxide.

CONCLUSION

We have reported research and development leading to a sol-gel process for optical fiber production. The key features of the process which make it commercially viable are: 1) the use of low cost fumed silica starting material; 2) the ability to cast, dry and sinter large unbroken bodies; 3) the ability to remove flaw inducing refractory particles by sol centrifugation and by treatment in a reactive atmosphere during firing; 4) a process to remove OH^- and transition metal ions to achieve needed purity and 5) the means of attaining and maintaining the needed dimensions of silica tubes to be used for overcladding optical fiber preforms.

ACKNOWLEDGMENT

The prototype process described has resulted from the efforts of many individuals. We thank D. A. Fleming, E. M. Rabinovich, E. M. Vogel, F. W. Walz, and M. F. Yan for their contribution to the early exploratory phase of the study. C. C. Bahr, S. Fleming, and P. L. Trevor provided material quality assessments. J. C. Alonzo, D. A. Mixon, R. M. Pafchek, and T. E. Stockert provided expertise in formulation, forming, purifying, and sintering of specimens. R. L. McAnally and M. Robel designed and prepared glass structures. L. J. Anthony and L. A. Psota-Kelty provided analytic measurements. We are indebted to H. C. Chandan for break analysis, D. H. Smith for overcladding experiments and to P. F. Glodis for measuring and evaluating loss characteristics. Many people at Lucent Technologies have provided valuable insight to this study. Of particular note are M. J. Cardillo, P. Dickinson, T. J. Miller, and M. I. Schwartz.

REFERENCES

1. Roy, R.: Aids in Hydrothermal Experimentation: II, Methods of Making Mixtures for Both "Dry" and "Wet" Phase Equilibrium Studies, Journal of American Ceramic Society **39**, 4, 149-6 (1956).

2. Yoshida, K.; Satoh, T.; Enomoto, N.; Yagi, T.; Hihara, H.; Oku, M.: Fabrication of Large Preforms for Low Loss Single Mode Optical Fibers by Hybridized Process. Glastech Bericht, (1996).

3. Clasen, R.: Preparation of High-Purity Silica Glasses by Sintering of Colloidal Particles. Glastech. Ber. **60**, 125-132, (1987).

4. Clasen, R.: Forming of compacts of Submicron Silica Particles by Electrophoretic Deposition. In: Hausner, H.; Messing, G. L.; Hirano, S. (eds.): Ceramic Powder Processing Science. Proceedings 2nd International Conference on Powder Processing Science, Berchtesgaden, Koln: Deutsche Keramische Gesellschaft. 1989, p. 633-640, (1988).

5. Bachmann, P. K.; Geittner, P.; Lydtin, H.; Romanowski; Thelen, M.: Preparation of Quartz Tubes by Centrifugal Deposition of Silica Particles. ECOC, Brighton, (1988).

6. MacChesney, J. B.; Johnson, D. W. Jr.; Bhandarkar, S.; Bohrer, M.; Fleming, J. W. ; Monberg, E. M. and Trevor, D. J.: Optical Fibers Using Sol-Gel Overcladding Tubes, Electronic Letters **33**, 18, 1573, (1997).

7. Chandross, E. A.; Johnson, D. W. Jr.; MacChesney, J. B.: U.S. Patent 5,379,364. Vitreous Silica Product via Sol-Gel Using Polymeric Additive, (1995).

8. Chandan, H.; Parker, R. D.; Kalish, D.: Fractography in Optical Fibers, in Fractography of Glass, R. C. Bradt and R. E. Tressler, eds., pp. 143-184, Plenum Press, New York, (1994).

9. Bhandarkar, S. D.: U.S. Patent 6,356, 447, (1994).

10. Iler, R.: The Chemistry of Silica. John Wiley and Sons, N.Y., (1979).

HISTORICAL DEVELOPMENT OF ABRASIVE GRAIN

D.D. Erickson, T.E. Wood, and W.P. Wood
3M Co., St. Paul, MN. 55144

Nowhere has the impact of sol-gel processing been more evident than in the field of abrasive particle technology. In 1981, 3M Company introduced the first sol-gel abrasive particle referred to as Cubitron™ grain in the Regal™ coated abrasive fibre disc product line. Products based on this sol-gel abrasive particle could grind three to five times more metal than any conventional fused alumina abrasive particle in an identical construction. Since 1981, sol gel abrasive particles have proliferated into vitrified and resin bonded grinding wheels, coated abrasive belts, sheets and discs and abrasive bristle discs. Today, these sol-gel abrasive particles have established themselves as the mineral of choice in a myriad of grinding applications. In fact, many operators in plants ranging from metal foundries to automotive assemblers and from paint shops to furniture manufacturers have now come to demand the level of exceptional grinding performance provided by abrasive systems based on sol-gel chemistry.

To fully appreciate the significance of the sol-gel abrasive particles it is helpful to examine both the natural and synthetic minerals which have been previously used as abrasive particles. The earliest abrasive minerals were harvested from the most common and most abundant minerals in the earth's crust, the quartz family. Called "crystal" by the ancient Greeks, quartz in the form of sandstone, loose sand and flint was used in the surfacing of stones, wood and metal and in the grinding of grain and the comminution of lime. The ancient Greeks also developed the use of corundum (emery) and garnet which were superior to quartz in abrasive applications. Theophrastus referred to haematite,[1] now known as Jeweler's rouge, in 325 BC, although it is unknown at what point this material began to be widely used as a polishing agent for jewelry. Diamond was used as a polishing powder in India by 800 BC[2] and its extraordinary hardness is noted in ancient Hindu proverbs[3] and in the Bible.[4]

Shark skin was another material utilized as sandpaper by many early cultures. Dermal denticles known as placoid scales grow on the surface of the skin of the shark and give the skin its abrasive properties. The uniform placement of the abrasive scales on this natural abrasive surface is known to provide a very uniform finish after polishing.

Until the late 19th century, abrasive particle technology remained relatively stagnant. In fact, when the first sandpaper was produced in 1902 by the newly founded 3M Company, this new product was coated with natural corundum imported from Greece. Of course, the abrasive performance of the new sandpaper varied with changes in the quality of the natural corundum. Thus, this new technology, sandpaper, provided an excellent application for newly developed, synthetic minerals that had begun to emerge.

The synthetic minerals that would revolutionize the abrasives industry at that time were fused alumina and silicon carbide, a material not found in terrestrial minerals. The process to make

To the extent authorized under the laws of the United States of America, all copyright interests in this publication are the property of The American Ceramic Society. Any duplication, reproduction, or republication of this publication or any part thereof, without the express written consent of The American Ceramic Society or fee paid to the Copyright Clearance Center, is prohibited.

fused alumina, also known as synthetic emery or corundum, was patented by Werlein in France in 1893[5] and by Hasslacher in Germany in 1894.[6] This process was improved in 1895 by the use of transition aluminas as precursor materials[7] and in 1900 by controlled cooling of the melt.[8] C. M. Hall further discovered that the addition of iron borings to the charge allowed the facile removal of metallic impurities.[9] An improved furnace was designed by A. C. Higgins which involved the use of a water-cooled steel shell container which developed a solid, thin alumina coating that served to protect the steel from attack by the hot alumina.[10] A description of fused alumina production can be found in a review by Cichy.[11]

Silicon carbide, a material harder than alumina, was synthesized accidentally by Edward Acheson in 1891.[12] Although probably prepared earlier by Despretz[13] and Marsden,[14] Acheson is generally credited with the discovery of silicon carbide. Acheson was trying to recrystallize carbon from a melt of aluminum silicate when he isolated blue crystals which were extremely hard. Because he imagined that this material was a compound of aluminum and carbon, he gave it the name *carborundum* from a combination of 'carbon' and 'corundum' (the common name for alpha alumina). This compound, however, proved to be a covalent carbide of silicon which was surprisingly oxidation resistant and extremely refractory . Silicon carbide is produced by the reduction of silica by carbon at temperatures in excess of 2000°C. Acheson further developed furnacing technology to manufacture the material which is still widely in use today.

For the next 70 years various forms of fused alumina and silicon carbide constituted the state of the art of high performance abrasives. Aluminous abrasives were used to grind steels due to the high melting point, high hardness at elevated temperature and low reactivity of alumina with the metal workpiece. Silicon carbide was used on lower temperature metals such as cast iron and non-metallic materials such as wood and paint. During this period the industrial use of diamond as an abrasive began in the 1930s and synthetic diamond was first produced in 1955.[15] Cubic boron nitride (CBN) was also synthesized during this period, and while both CBN and diamond were far superior abrasives in many applications, their usage remained small due to their extremely high cost.

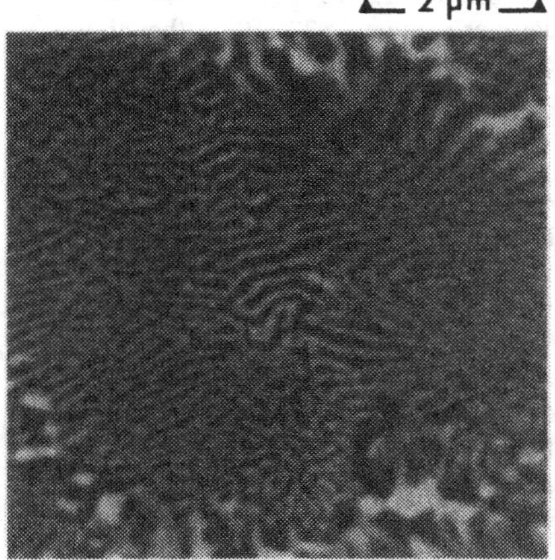

Figure 1. Backscattered electron image of fused alumina-zirconia abrasive grain.

In 1973 a new fused grain was introduced by Norton Company which raised the level of abrasive particle performance.[16] This grain was similar to fused alumina in that it was derived from a melt process. However, this material was based on an eutectic alloy melt of alumina and zirconia. Rapid solidification of the melt resulted in the growth of small, less than 50 microns in diameter, interpenetrating dendrites of alumina and zirconia (Figure 1). This rapid cooling also quenched the zirconia into the metastable tetragonal phase so that martinsitic toughening further improved the abrasive properties of this grain.

This new alumina-zirconia grain significantly outperformed the single crystal fused grains in most applications and clearly demonstrated the performance advantages which could be realized by control of the abrasive particle crystalline form and microstructure.

It is now widely recognized that in polycrystalline ceramic bodies, microhardness and fracture toughness generally increase with decreasing size of the individual crystallites. Simply put, abrasive particles which have high hardness at the temperature of grinding and are resistant to fracture will generally work longer and more efficiently. Control of the microstructure to produce crystal sizes which are more uniform, defect free and of a small size is thus often associated with increased abrasive performance and longer performance lifetime.

The response at 3M to this new alumina-zirconia abrasive was to establish a research program to develop new, high performance abrasives which were tough and also very fine grained. The idea to use sol-gel processing to meet this objective was first suggested by Hal Sowman of 3M. In earlier projects Sowman's group had successfully developed a variety of ceramic fiber and particle systems and nuclear fuel technologies using sol gel processing. [17] The abrasive area provided another potential application and new challenges for this technology.

Research initially centered on using a commercially available boehmite (AlOOH) as the source of the alumina phase. Boehmite has the distinction of being extremely dispersible in water and is also amenable to sol-gel processing. The boehmite powder of choice is derived from the hydrolysis of aluminum alkoxide and is a byproduct of the Ziegler process for the production of long chain alcohols. As such, this alumina is relatively pure, with titania being the major impurity at 0.2% or less. The boehmite crystallites are hexagonal platelets less than 15 nm in size which are agglomerated into particles generally less than 100 nm in diameter. By dispersing the boehmite powder in acidified water, a well-dispersed colloid is formed by the electrostatic stabilization of the crystallites. This dispersion could be dried to form a rigid xerogel which could be crushed to a desired size and fired into abrasive particles. Firing boehmite-derived gels, however, resulted in porous, vermicular microstructures which were friable and weak. The materials could be densified by firing at very high temperatures, but this resulted in exaggerated grain growth and loss of strength. Therefore, initial research was focused on developing grain growth inhibitors and sintering agents in order to enable the production of dense alumina abrasive particles from boehmite at lower firing temperatures.

In a general process for making a sol-gel particle, a sol is first made by dispersing boehmite powder in acidified water with any undispersed material being removed, typically by a centrifuge. Next a metal oxide modifier, typically in the form of a soluble metal salt is mixed into the sol. As the metal salt is added to the sol, gelation occurs and a stiff gel is formed. Afterwards, this gel is dried, crushed and screened to the appropriate size distribution. There are several advantages to crushing the dried gel rather than the sintered product. First, crushing the dried gel requires considerably less energy and causes less wear on machinery than crushing the fired product. Second, the very fine particles of crushed gel can be redispersed into a sol and thus recycled. After crushing and screening, the resulting particles are calcined at 600° to 900° C and then sintered at a temperature between 1200° to 1600° C. The sintering step results in a dense ceramic body that is screened to the appropriate particle size and is ready to be used as an abrasive particle.

Initially, it was discovered that certain additions of zirconia or magnesia in the form of precursor salt solutions to the boehmite sol produced gels that could be fired to generate dense alumina ceramic bodies at low firing temperatures. [18] The resulting abrasive grain had many of the desirable physical properties, but most importantly, in grinding tests the amount of metal removed was equal to, or often greater than that of the fused alumina-zirconia abrasive grain. In 1981 the first sol-gel abrasive grain, referred to as MBM Cubitron™ grain, was incorporated into a coated abrasive fibre disc and was introduced to the general metal working market. Although this Cubitron™ grain containing fibre disc had outstanding grinding performance, acceptance of product was initially slow because of its high costs when compared to the fused alumina-zirconia abrasive grain. Even today, sol-gel processing costs remain high in comparison to fused abrasive grain. Through the passage of time, though, end user customers have come to embrace the high performance that the sol-gel abrasive grains offered. As a result, the growth of sol-gel abrasive products continues today, seventeen years after the first commercial introduction.

The constituents of the original, commercial sol-gel abrasive particle comprised Al_2O_3 and MgO. During the sintering step, the magnesia and a portion of the alumina react to form a spinel phase ($MgO \cdot Al_2O_3$). The remainder of the alumina transforms sequentially from boehmite to gamma, delta and finally theta alumina before converting to alpha alumina at about 1220°C. The alpha alumina grows in the theta alumina in a cellular or vermicular growth habit as described by Dynys and Halloran,[19] and shown in Figure 2. The spinel, present as 50 to 200 nanometer precipitates within the cells, contributes to the fracture toughness of the ceramic body (3.8 $MPam^{0.5}$), but the spinel phase also slightly lowers the hardness of the particle relative to pure alpha alumina (17 GPa versus 20-21 GPa).

Figure 2. SEM micrograph showing cellular growth of alumina in Cubitron™ grain.

The original sol-gel abrasive particle, the MBM Cubitron™ grain, had found wide acceptance in metalworking industries, especially for grinding mild and tool steels, but fused alumina-zirconia had continued to dominate the market for grinding stainless steel and exotic alloys. To fill this gap, sol-gel development work continued at 3M Co. and a new mineral was developed which was found to be extremely effective in grinding stainless steel. This new abrasive particle, denoted as MYM Cubitron™ grain, contained yttrium oxide which reacted with the alumina during sintering to form approximately 60 nm precipitates of yttrium aluminum garnet (YAG).[20] The YAG limited the cellular growth of the alumina to 3 micron cells containing submicron alpha alumina crystallites. While this yttria modified sol-gel abrasive particle performed well in most applications, it was discovered that its performance was dramatically enhanced when it was used in combination with a grinding aid. Grinding aids are generally halogen containing compounds which are applied as a thin coating to the top surface of a coated abrasive article or impregnated into the porosity of a grinding wheel. During use, the temperatures generated by grinding cause the grinding aid to decompose and the resulting decomposition products react with the freshly abraded metal surface. This reaction with the metal surface prevents the metal from adhering to the mineral surface and failing by being coated with metal in a condition known as "capping." Although this abrasive particle was expensive to manufacture due to the high yttria content, when this abrasive particle was incorporated into coated abrasives the resulting enhanced grinding performance on stainless steel provided substantial value to the industrial customer.

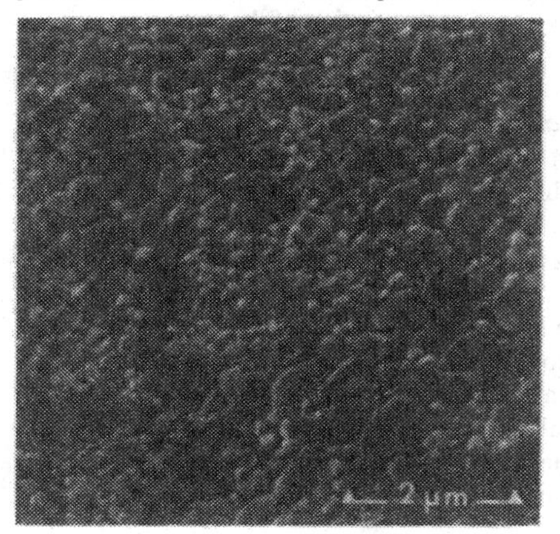

Figure 3. SEM micrograph of Cubitron™ grain nucleated with alpha alumina seed particles.

By 1984, researchers working independently at Penn State,[21] Carborundum,[22] Norton[23] and 3M[24] had found that the addition of sub-micron particles of alpha alumina to the boehmite sol would drastically change the growth of alpha alumina from a cellular growth habit to that of discrete equiaxed crystals of random orientation as shown in Figure 3. Furthermore, the transformation temperature from theta to alpha alumina was lowered by more than 100°C.[21] It is thought[21] that the sub-micron alpha-alumina seed particles provide a template for the transitional alumina to nucleate to alpha-alumina.

Although dense, sub-micron alpha alumina had been previously prepared by methods such as hot-pressing,[25] this sol-gel process enabled a level of microstructural control that had previously been unattainable. This phenomenon of "seeding" was to be widely described in the technical literature over the next several years.[26] While the addition of alpha alumina seeds to Cubitron™ grain did not result in an improvement of the abrasive properties, it did offer significant processing advantages such as a lower sintering temperature. In 1986 a seeded version of Cubitron™ grain called MBMS which also contained magnesium spinel was commercialized and replaced the original MBM. As MBMS did not offer any advantage over MBM in the grinding of stainless steel, the non-seeded yttria containing mineral continued to be produced.

Other materials that are isostructural with alpha-alumina were also found to be effective nucleating agents for the transformation to alpha alumina. Alpha iron oxide differs by about 5% in lattice parameter dimensions when compared to alpha alumina. Consequently, alpha iron oxide can also function as a nucleating agent for the transformation to alpha alumina. With 3M's considerable expertise in the production of magnetic pigments, it was found that greater control of the iron oxide particle size and distribution resulted in improved processing and performance over that obtained with alpha alumina seeds. In 1987, an alpha iron oxide precursor replaced alpha alumina as the seed particles and a new mineral, MLM Cubitron™ grain, replaced MBMS.

McArdle *et al*[27] were to later confirm that the growth of alpha alumina on large alpha ferric oxide seed crystals occurs by solid phase epitaxy. However, researchers at 3M and the University of Minnesota using sub-micron iron oxide particles to nucleate the growth of alpha-alumina from transitional alumina noted several anomalies which suggested that when very small particles of iron oxide or iron oxy-hydroxide were used, such as is used in the MLM Cubitron™ grains, the chemical interaction of the iron and alumina began to play a more important role. First, the x-ray diffraction pattern for an iron oxide nucleated sol-gel ceramic showed alpha iron oxide to be present when fired to 600°C, but absent when fired to 900°C, a temperature below the alpha alumina transformation temperature. Second, iron oxide seed particles which can be observed in the matrix of gamma alumina when the grain is heated to 600°C, disappear in the matrix of theta alumina when heated to 900°C. Both observations suggest that dissolution of the very small iron oxide particles into the transition alumina to form a solid solution is occurring, and that the formation of alpha alumina is not simply the direct result of heterogeneous nucleation off the alpha iron oxide seed crystal. Interestingly, Lange *et al*[28] recently described a partitioning mechanism for aluminum -iron oxide solutions which may also play a role in the nucleation of the alpha phase from the aluminum-iron solid solutions that are formed in these systems.

Chromia was also found to be a successful nucleating agent.[29] Surprisingly, for the chromia to effectively nucleate the transformation to alpha alumina it was found that it is advantageous for at least a portion of the surface chromium to initially be in a higher oxidation state to establish interaction with the aluminous phase. The material could then be processed in a reducing atmosphere to reduce the chromium back to the +3 state. Dense, fine microstructured alumina-chromia abrasive minerals containing equiaxed crystals which are less than 0.5 micron were prepared in this manner. The complex behavior of the iron-based and the chromium-based nucleating agents well illustrates the chemical nature of this nucleation phenomenon.

By 1990 a new abrasive particle was developed for grinding stainless steel and exotic alloys. This new sol-gel abrasive particle, designated 321 Cubitron™ grain, contained lower cost rare earth oxides, namely, lanthanum and neodymium.[30] These rare earth metal ions react with the magnesia and alumina during firing to form a hexagonal aluminate with a magnetoplumbite crystal structure. This magnetoplumbite phase was shown to exist as platelets about 50 nm thick and up to 1 micron in diameter. These platelets were interlaced within a 3 to 5 micron cellular growth of submicron alpha alumina crystallites (Figure 4).

Figure 4. SEM micrograph showing platelet growth without iron oxide seed additions.

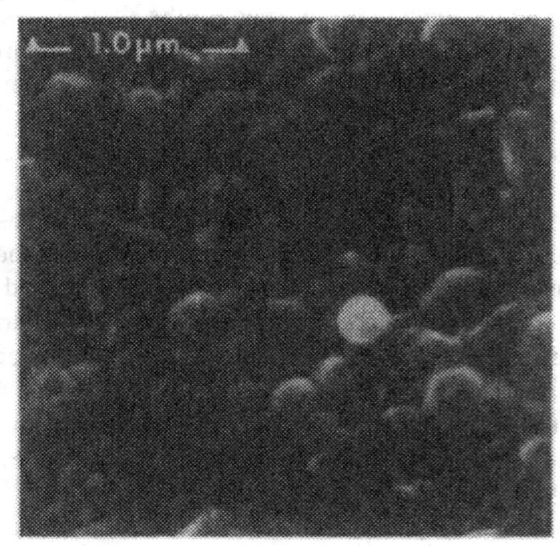

Figure 5. SEM micrograph showing equiaxed particle growth with iron oxide seed added.

If iron oxide seed particles[31] are added to this composition, then the magnetoplumbite phase grows not as platelets, but as submicron, equiaxed particles between submicron alpha alumina crystals

(Figure 5). This is due to the reduction in the alpha alumina transformation temperature in the seeded composition which requires the magnetoplumbite phase to grow in a matrix of fully formed alpha alumina. In the case of the non-seeded composition, the magnetoplumbite forms simultaneously with alpha alumina and thus has the opportunity to form in the desired morphology before the surrounding alpha alumina matrix has fully formed. The Vickers hardness for the seeded composition can be as high as 22.0 GPa whereas the hardness of the non-seeded composition is 19.6 GPa. However, the non-seeded grain has a fracture toughness of 3.7 $MPam^{0.5}$ versus a fracture toughness of about 3.3 $MPam^{0.5}$ for the seeded grain which demonstrates the contribution to fracture toughness of the platelet morphology. As a result, the non-seeded abrasive grain containing rare earth oxide modifiers was commercialized in 1990. This grain is still in production and is used in both 3M's Regalloy™ and Regalite™ lines of coated abrasives as well as being sold world-wide as 321 Cubitron™ loose grain to manufacturers of both vitreous and resin-bonded grinding wheels.

In addition to increasing fracture toughness, the platelets serve another beneficial function in the abrasive particle with respect to its application in vitreous bonded wheels. Vitreous bonded wheels are abrasive wheels that contain a vitreous binder which is typically matured and densified at temperatures ranging from about 900° to 1300° C. If the abrasive particle used in the grinding wheel is reactive or not sufficiently refractory, the higher temperature vitreous binders may cause excessive growth of the crystals in the polycrystalline abrasive particles. If this excessive crystal growth does occur, the abrasive particle strength and its grinding performance will be reduced. Thus, in some instances lower temperature vitreous binders are required to minimize any crystal growth. The platelets in the 321 Cubitron™ grain are believed to pin the alumina crystallites such that at relatively higher temperatures there is no significant alumina crystal growth. Thus 321 Cubitron™ grain can be used with great success with higher temperature vitreous binding systems.

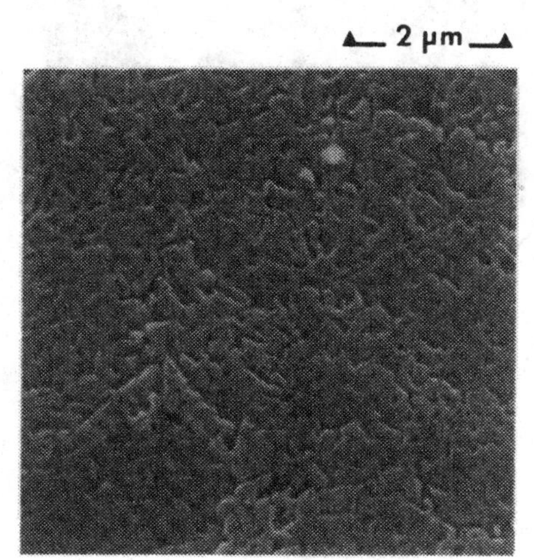

Figure 6. SEM micrograph showing attack of magnetoplumbite phase by grinding fluid.

The grinding performance of the 321 Cubitron™ grain is also enhanced with the use of halogen grinding aids, but this enhancement is greater than for other sol-gel grains which do not contain the platelet morphology. It is believed that in addition to reacting with the freshly abraded metal surface to reduce the amount of capping, the halogen compound also attacks the magnetoplumbite platelet phase. In Figure 6, a polished surface of 321 Cubitron™ grain is shown which has been hydrothermally treated in a halogenated grinding fluid to simulate conditions encountered during grinding.

The magnetoplumbite phase has clearly been attacked by the halogen leaving the surrounding alpha alumina matrix intact. This attack is believed to weaken the alumina surface causing a controlled erosion of the surface which also removes any metal adhering to the surface.

One grain which has not been commercialized but is of special interest contains both manganese and lanthanum oxides as modifiers to the alpha alumina matrix.[32] Manganese oxide is known to lower the alpha alumina transformation temperature.[33] Interestingly, the hexagonal form of Mn_2O_3 is reported to convert to the cubic alpha modification by 565°C.[34] Despite this lower transformation temperature, the resulting microstructure retains a vermicular growth habit characteristic of the non-seeded alumina containing 6% MnO as shown in Figure 7. On the other hand, lanthanum oxide is reported to stabilize the structure of the transitional aluminas and thus retard the formation of alpha alumina.[35] With the addition of 3% La_2O_3 and 6% MnO to sol gel alumina, a dense, non-cellular growth of submicron alumina is obtained with equiaxed magnetoplumbite and spinel phases between the triple points (Figure 8). By increasing the La_2O_3 content to 5% with 6% MnO, the alpha alumina transformation temperature is further increased, and a cellular growth habit of submicron alpha alumina is obtained with secondary phases present within the cells and between cell boundaries (Figure 9).

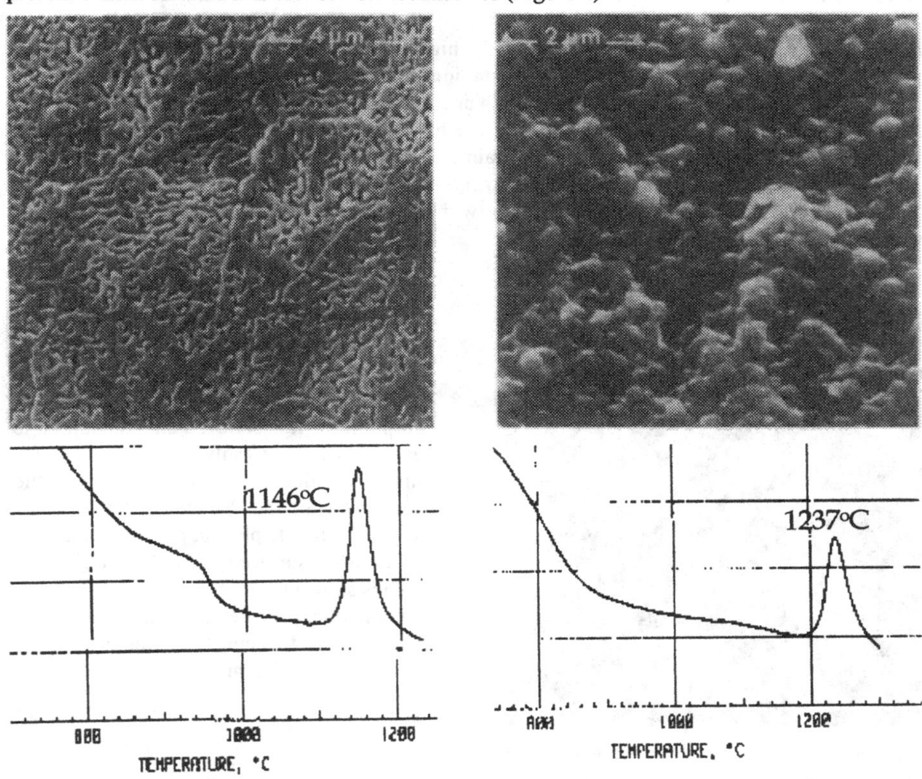

Figure 7. SEM micrograph and DTA of vermicular growth with 6% MnO, 1% La_2O_3

Figure 8. SEM micrograph and DTA of non-cellular growth with 6% MnO, 3% La_2O_3.

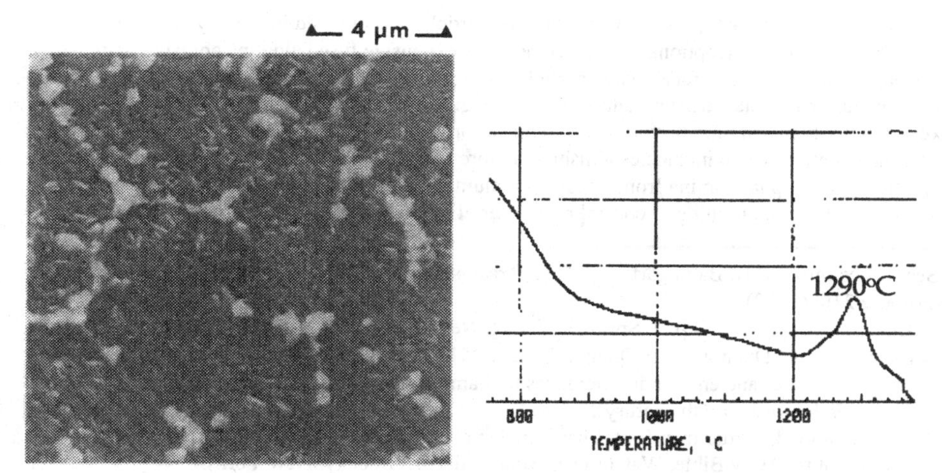

Figure 9. SEM micrograph and DTA of cellular growth with 6% MnO and 5% La_2O_3.

Non-oxide abrasive grains can also be produced by a modification to the sol gel process[36]. One such grain is a composite of TiN dispersed in a matrix of either AlN or gamma aluminum oxynitride (γ-AlON) which is made by the addition of TiO_2 and carbon to a seeded boehmite sol. By firing to high temperature in a nitrogen atmosphere, the nitrides of titanium and aluminum are formed by carbothermal reduction. By lowering the carbon content of the sol, the amount of Al_2O_3 which converts to AlN can be limited. In this case, AlON is then formed by reaction between the AlN and the residual Al_2O_3. The TiN exists as submicron, equiaxed particles uniformly distributed between the 1 to 2 micron grains of the γ-AlON.

The use of chemical modifiers to form a secondary phase within the alpha alumina matrix thus permits the development of many unique microstructures which exhibit different fracture behaviors. As a result, the abrasive particle can be tailored to be more friable or less friable depending upon the target grinding application. For example, for high pressure metal grinding of tool steel, a tough abrasive particle was typically desired to be able to withstand high grinding forces. To achieve this, magnesium nitrate can be added to the alumina sol and during sintering the magnesia reacted with alumina to form a spinel phase. The presence of this spinel phase increased the overall fracture toughness of the abrasive particle. Alternatively, the formation of the magnetoplumbite phase with a platelet morphology also increases fracture toughness. For low pressure grinding of wood or paint, a more friable abrasive particle is sometimes preferred. This can be achieved by removing the secondary phase entirely from the alumina matrix, or by altering the form of the secondary phase as from a platelet morphology to equiaxed particles. As a result, the friable abrasive particles can continuously break down at low pressures to provide consistently high cutting rates and long life. If this breakdown at these low pressures did not occur, then the abrasive particle tended to dull and the cutting of the wood or paint surface would rapidly fall off.

After 18 years, the sol-gel processing of abrasive particles has evolved into many abrasive products which show exceptional value over the less expensive fused alumina abrasive particles. A strong advantage of sol-gel abrasive particles is the ability to introduce chemical changes which alter the alumina crystal structure and enable the engineering of the abrasive particle to provide exceptionally high grinding performance compositions for different grinding applications. Many challenges remain which include establishing a more complete understanding of the growth and densification of alpha alumina from transitional alumina, identifying new and better performing compositions, and continuing to control raw material and processing costs to remain competitive.

---

[1] See for example A. H. Baumgärtner, "Theophrastus von den Steinen aus dem Griechischen", Nürnberg 210, (1770).
[2] L. Coes, Jr., "Abrasives," p. 2, Springer-Verlag, New York, 1971.
[3] Hindu proverbs "Diamond cuts Diamond," and "The heart of a magnate is harder than diamond. For other ancient Hindu references to diamond see also *The Hindu Vedas* (1100 - 1200 BCE), and *Brhatsanhita* (6th century).
[4] "Like a diamond, harder than flint, I have made your forehead..." Ezekiel 3:9 (New World Translation of the Holy Bible, Watchtower Bible and Tract Society, New York).
[5] I. Werlein, "Process for Hardening Aluminum Materials by Electric Fusion," French Patent 233,996.
[6] F. Hasslacher, "Method of Converting Natural Emery into Iron and Water-Free Fused Alumina," German Patent 85,021.
[7] I. Werlein, "Process for Hardening Aluminous Materials by Electric Fusion," addition to French Patent 233,996.
[8] C. F. Jacobs, "Process of Manufacturing Abrasive Minerals," U. S. Patent 659,926.
[9] C. M. Hall, "Process of Purifying Bauxite," U. S. Patent 677,207.
[10] A. C. Higgins, "Electric Furnace," U. S. Patent 775,654.
[11] P. Cichy, "Fused Alumina Production," Paper No. EFC-7, The Metallurgical Society of AIME, New York, NY, 10017.
[12] E. G. Acheson, U. S. Patent 492,767; U. S. Patent 527,826; U. S. Patent 650,291; U. S. Patent 615,648; U. S. Patents 718,891, 718,892, 722,792, 722,793, and 723,631.
[13] C. M. Despretz, *Compt. Rend.*, **29**, 720 (1849).
[14] R. S. Marsden, *Proc. Roy. Soc.*, **11**, 20 (1881).
[15] F. P. Bundy, H. T. Hall, H. M. Strong, and R. H. Wentorf, Jr., *Nature*, **176**, 51 (1955).
[16] R. A. Rowse and G. R. Watson, U. S. Patent 3,891,408; J. R. Quinan and J. E. Patchett, U. S. Patent 3,893,826.
[17] H. G. Sowman, U. S. Patent 3,795,524,; U. S. Patent 3,709,706,; U. S. Patent 3,793,041,; U. S. Patent 3,916,584,; U. S. Patent 4,047,965,; U. S. Patent 4,125,406.
[18] M. A. Leitheiser and H. G. Sowman, U. S. Patent 4,314,827.
[19] F. W. Dynys and J. W. Halloran, *J. Am. Ceram. Soc.*, **65**[9], 442 (1982).
[20] L.D. Monroe and W.P. Wood, U.S. Patent 4,770,671.
[21] M. Kumagai and G. L. Messing, *Comm. Am. Ceram. Soc.*, C-230 (1984); M. Kumagai and G. L. Messing, *J. Am. Ceram. Soc.*, **68**[9], 500 (1985).
[22] A. P. Gerk, U. S. Patent 4,574,003.
[23] T. E. Cottringer and R. H. van de Merwe, U. S. Patent 4,623,364.
[24] M. G. Schwabel, U. S. Patent 4,744,802.

[25] See for example: C. J. P. Steiner, R. M. Spriggs and D. P. H. Hasselman, *J. Am. Ceram. Soc. - Discussion and Notes*, 1972, **55**[2], 115; J. Bugosh, U. S. Patent 3,141,786 (see figure 6); L. Coes, Jr., U. S. Patent 3,909,991; H. W. Stetson, W. J. Gyurk, U. S. Patent 3,698,923.
[26] See for example: G. L. Messing and M. Kumagai, *Amer. Ceram. Soc. Bull.*, **73**[10], 88 (1994); A. V. Prasadarao, U. Selvaraj, S. Komarneni, A. S. Bhalla and R. Roy, *J. Am. Ceram. Soc.*, **75**[6], 1529 (1992); T. Y. Tseng, Y. Y. Kuo, and Y. L. Lin, *J. Mater. Sci. Lett.*, **8**[11], 1274 (1989); H. Yaparlar and H. Hausner, *Ceram. Trans.*, Vol. 22, Ed. S. Hirana, G. L. Messing, and H. Hausner, The American Ceramic Society, Inc., Westerville, Ohio.
[27] J. L. McArdle and G. L. Messing, *Adv. Ceram. Mater.*, **3**[4], 387 (1989); J. L. McArdle, G. L. Messing, L. A. Tietz, and C. B. Carter, *J. Am. Ceram. Soc.*, **72**[5], 864 (1989).
[28] A.D. Poli, F.F. Lange, and C.G. Levi, *J. Am. Ceram. Socl*, **79**[7], 1745 (1996).
[29] T. E. Wood, U. S. Patent 5,219,806.
[30] W.P. Wood, L.D. Monroe, and S.L. Conwell, U.S. Patent 4,881,951
[31] D.D. Erickson and W.P. Wood, *Ceram. Trans.*, **46**, 463 (1994). Use of the terms "seeding" or "seed" is not meant to imply that the particles in question are added to induce alpha phase formation via epitaxy. It is clear that in many cases enhancement of alpha alumina transformation is not occurring via simple epitaxy.
[32] W.P. Wood and H.A. Larmie, U.S. Patent 5,690,707.
[33] Y. Wakao and T. Hibino, *Reports of Nagoya Industrial Research Institute*, **11**[9], 588 (1962).
[34] J. A. Lee, C. E. Newnham, F. S. Stone, and F. L. Tye, *J. Solid State Chem.*, **31**, 81 (1980).
[35] J. S. Church and N. W. Cant, *Applied Catalysis A: General*, **101**, 105 (1993).
[36] J.P. Mathers, T. Forrester, and W.P. Wood, U.S. Patents 4,788,167; 4,855,264, and 4,957,886; *Ceramic Bulletin*, **68**[7], 1330 (1989).

Electroceramic Films by Sol-Gel Process

LOW TEMPERATURE PROCESSING OF SOL-GEL DERIVED $Pb(Zr_{1-x}Ti_x)O_3$ THIN FILMS

Yoon J. Song, S. Tirumala, and Seshu B. Desu
Department of Materials Science and Engineering,
Virginia Tech, Blacksburg, VA 24061-0237

ABSTRACT

In this study, we reported low temperature processing and properties of $Pb(Zr_{1-x}Ti_x)O_3$ (x = 0.6, 0.7, and 0.8) films by a modified sol-gel technique. Using the sol-gel processing, it was possible to prepare highly (111) oriented PZT films on $Pt/Ti/SiO_2/Si$ at a low annealing temperature of 500 °C. The structural and electrical properties of the PZT films processed at 500 °C were investigated as a function of Ti content. Well-saturated hysteresis loops were observed for the Ti-rich PZT films at 500 °C. As the Ti content varied from 0.6 to 0.8, the value of remanent polarization (P_r) increased from 5 to 14 $\mu C/cm^2$ at an applied voltage of 5 V, because of enhanced crystallization of the PZT20/80 films at 500 °C. As the Ti content increased, the leakage current density of the PZT films were found to increase at the annealing temperature of 500 °C.

INTRODUCTION

In recent decades, ferroelectric random access memory (FRAM) device has attracted great attention because of relatively low operating voltage, excellent radiation hardness and compatibility with existing silicon VLSI technology.[1,2] In particular, $Pb(Zr_{1-x}Ti_x)O_3$ (PZT) thin films are regarded as a strong candidate for the FRAM devices due to excellent ferroelectric properties and relatively low processing temperature.

The basic unit of the current destructive readout (DRO) FRAM is 1 transistor and 1 capacitor,[3] where the PZT films are used as capacitor materials. As the memory density increases, the cell areas are so tremendously reduced that the capacitor is stacked on the top of the transistor rather than being placed next to the transistor, which is seen in low density memory devices. This stacked structure demands the use of the plug (polysilicon) in high density DRO FRAM to establish communication between the ferroelectric capacitor and the transistor. Typically, ferroelectric films are prepared at relatively high annealing temperature in strong

To the extent authorized under the laws of the United States of America, all copyright interests in this publication are the property of The American Ceramic Society. Any duplication, reproduction, or republication of this publication or any part thereof, without the express written consent of The American Ceramic Society or fee paid to the Copyright Clearance Center, is prohibited.

oxygen environment, which causes the polysilicon to be easily oxidized and form an insulating layer such as SiO_2. Since the insulating layer severely destroys the device performance, diffusion barrier layers are required to prevent the undesired detrimental layer. Titanium silicide and titanium nitride are currently recognized as most effective barrier layers due to their outstanding conductivity and thermal stability.[4,5] However, the diffusion barrier layers are stable only up to 550 °C, which is lower than the typical ferroelectric processing temperatures (>650 °C). After integrating the ferroelectric films into the polysilicon, the barrier layers lose their conductivity and become insulating. Therefore, in order to integrate the ferroelectric films into the existing diffusion barrier layers, low temperature processing (<550 °C) is strongly demanded.

In the sol-gel technique, the structural and electrical properties of the final films are strongly dependent on the nature of precursor solution, deposition conditions, compositional ratio and the substrate. In particular, the chemistry and homogeneity of PZT precursor solution have strong influence on the crystallization temperature of the PZT thin films. In our experiment, the PZT solution was molecularly modified by adding a chemical agent, i. e. acetic acid. The chelating agents help in improving the homogeneity of precursor solution by generating highly dense polymeric network with high mixing rate. In addition, the low temperature processing can be assisted by selecting a suitable substrate for providing the nucleation sites for the perovskite phase, as it can reduce the activation energy for crystallization. Since the perovskite phase formation of sol-gel derived PZT films is controlled by nucleation[6], the role of the heterogeneous nucleation sites is important. Therefore, we selected $Pt/Ti/SiO_2/Si$ substrates which give the nucleation sites for perovskite phase due to good lattice matching.[7] Another approach to lower processing temperature is to inhibit the nucleation and crystallization of pyrochlore phase at a low temperature by rapidly annealing the samples without going through the pyrolysis step. In the present work, the samples were dried on a hot plate at 150 °C and directly annealed at 500 °C for crystallization. As the perovskite phase formation temperatures decrease as a function of Ti content, the Ti-rich PZT films were primarily investigated for the low temperature processing. In this report, we prepared PZT films with various composition (x = 0.6, 0.7, and 0.8) by a modified sol-gel technique.

EXPERIMENTAL PROCEDURES

In this experiment, we adopted and modified the procedure suggested by Yi.[8] 0.4 M $Pb(Zr_{1-x}Ti_x)O_3$ solutions (x = 0.60, 0.70, and 0.80) were prepared by using Zr-n-propoxide, Ti-isopropoxide, and lead-acetate as starting materials, and n-propanol as solvent. At first, Zr-n-propoxide was mixed with Ti-isopropoxide in appropriate molar ratio, and then acetic acid was added as a chemical modifier in the PZT precursor solution. Even small amount of acetic acid increases drastically gelation time, indicating that the solution is very stable under ambient atmosphere. Approximately 10 ml n-propanol was used as a solvent, and Pb acetate was put into the precursor solution. 10% lead excess was used to compensate for the lead loss during the heat-treatment process. The solution was then heated up to 85 °C for dissolving the lead more efficiently. After the lead acetate trihydrate was completely dissolved in the solution, acetic acid was added into the precursor solution as a chemical modifier and generated new type of precursor solution, which is resistive to hydrolysis and condensation. The PZT solution was molecularly modified by optimizing molar ratio of acetic acid and Ti-isopropoxide ([Acet]/[Ti]) before adding water. The effects of the chemical modifier on structural and electrical properties of final films were investigated by synthesizing five different solutions (R=[Acet]/[Ti]=2.5, 13, 25, 75, 250). After the chemical modification, water was dropped in the solution for hydrolysis and condensation. The precursor solution was mixed for 24 hour to enhance the mixing of modified precursors and produce more homogeneous solution. The precursor solution was then deposited on $Pt/Ti/SiO_2/Si$ substrates by spin-coating technique. The as-deposited layer was dried on a hot plate in air at 150 °C to evaporate water, alcohol and organics. The drying temperature was determined by the thermogravimetric analysis (TGA) results which showed that at 150 °C most of the alcohol, acetic acid, and water were removed from the sample. The spin-coating and drying process was repeated three times, and then the PZT films were finally pyrolyzed and annealed at various temperatures ranging from 450 to 600 °C. The PZT films were annealed in a pre-heated furnace under an oxygen atmosphere. After annealing, the samples were taken out and cooled under room temperature condition. The thickness of the post-annealed film with three layers, as measured by ellipsometry, was 0.25 μm.

The structure development of the films was investigated by x-ray diffraction (XRD) patterns, which were recorded on a Scintag XDS 2000 diffractometer using CuKα radiation at 40 kV. The ferroelectric measurements were carried out at room temperature by using a standardized RT66A ferroelectric test system operating in a Virtual-Ground mode. For electrical measurements, Pt electrodes were deposited on the top surface of the films through a shadow mask by RF sputtering.

RESULTS

The effect of acetic acid on the crystallinity of PZT films was investigated by varying a molar ratio of acetic acid and titanium isopropoxide ([Acet]/[Ti]) in the PZT solution. It was found that the best results, structural and electrical, were obtained for the PZT films prepared from the solutions with [Acet]/[Ti] = 25. Using the modified sol-gel solution with the optimal molar ratio, it was possible to prepare the PZT films with various compositions (x=0.6, 0.7, and 0.8) at a low annealing temperature of 500 °C. It was observed that the perovskite phase formation temperature of the PZT films decreased as the Ti content increased. Figure 1 shows the XRD patterns of Ti-rich PZT films annealed at 500 °C. Highly (111) oriented perovskite phase was observed without any pyrochlore phase for the Ti-rich PZT films at 500 °C. It was reported that the (111) grain orientation originated from intermetallic compound (Pt_3Ti) formed on $Pt/Ti/SiO_2/Si$ substrates.[7] The (111) oriented PZT was preferentially formed due to low interfacial energy which results from the fact that lattice parameter of intermetallic compound (Pt_3Ti) matched closely with that of (111) PZT. The lattice matching gives nucleation sites for PZT films to grow preferentially along (111) direction at a low annealing temperature of 500 °C. The argument that (111) orientation is generated from the lattice matching between Pt_3Ti and (111) PZT was supported by depositing PZT films on different substrates with no Ti underlayer such as $Pt/TiO_2/SiO_2/Si$ and Pt-Rh/Pt-RhO_x/Pt-Rh. Pyrochlore phase and poor electrical properties were observed for PZT films deposited on $Pt/TiO_2/SiO_2/Si$ and Pt-RhO_x/Pt-Rh/Pt-RhO_x at 500 °C, because these substrates do not provide heterogeneous nucleation sites for PZT perovskite phase, and thus the activation energy for perovskite phase is not reduced.

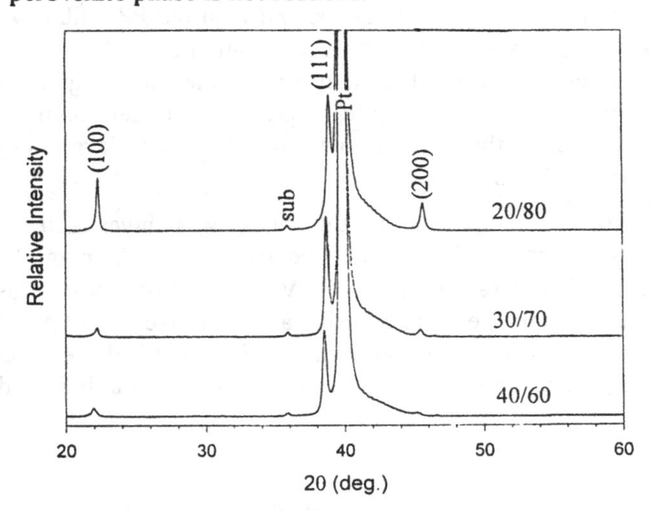

Fig. 1 X-ray diffraction (XRD) patterns of PZT (x=0.60, 0.70, and 0.8) films at an annealing temperature of 500 °C.

Figure 2 shows the P-E hysteresis loops of the PZT films with various Ti compositions. Well-defined hysteresis loops were observed for the Ti-rich PZT films at a low annealing temperature of 500 °C. It was found that the remanent polarization increased as the Ti content increased, which might be attributed to enhanced crystallization of perovskite phase for the Ti-rich films at the low annealing temperature. The value of remanent polarization depends on the extent of crystallization of perovskite phase.

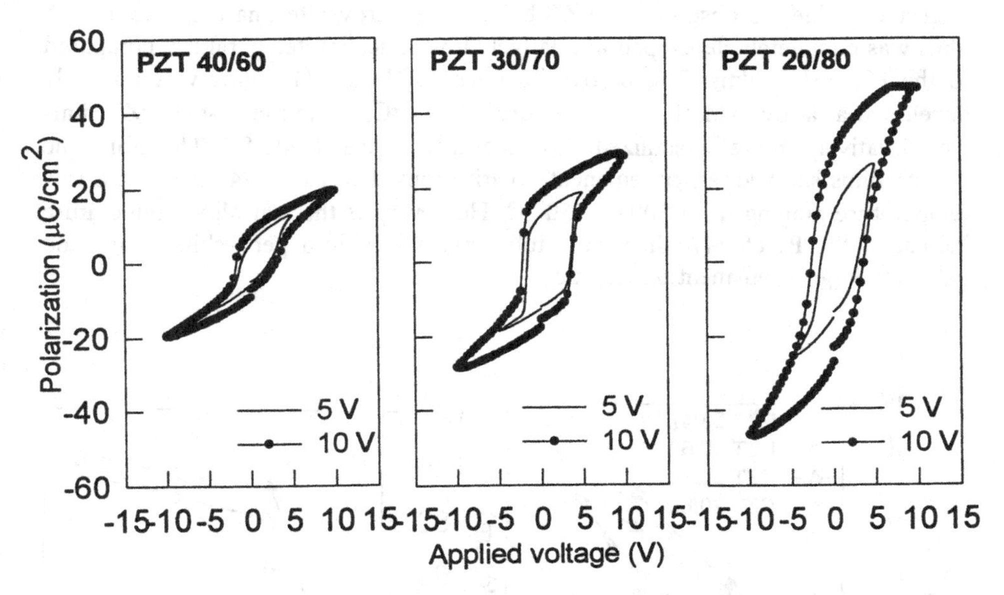

Fig. 2. Hysteresis loops of the PZT40/60, 30/70, and 20/80 films at 500 °C.

Figure 3 shows the dependence of P_r and E_c on the annealing temperatures. For comparison, P_r and E_c values of PZT53/47 films were included. It has been reported that remanent polarization is the highest for the PZT53/47 and decreases as Ti content increases. The reason of the largest remanent polarization for the PZT53/47 is that most domain switching orientations are possible at the morphotropic phase boundary (MPB) region. However, at the low annealing temperature region between 500 °C and 550 °C, largest P_r was observed for the PZT20/80 films, which showed lower phase formation temperature than other PZT films. As Ti content increased from 0.47 to 0.8, the remanent polarization increased from 1 to 14 $\mu C/cm^2$ for the PZT films at 500 °C. At an annealing temperature of 500 °C, the PZT20/80 films were almost fully crystallized into perovskite phase due to the low phase formation temperature. On the other hand, the perovskite phase of PZT53/47 films were not sufficiently developed at 500 °C to exhibit well-defined P-E hysteresis loops. As the annealing temperature increased from 500 to 550 °C, the value of P_r for PZT53/47 films increased from 1 to 12 $\mu C/cm^2$ due to the appreciable perovskite phase formation. At 600 °C, the PZT53/47 films were completely crystallized into perovskite phase and showed largest Pr value, as observed in PZT bulks. The perovskite phase of the Ti-rich films was completely developed at relatively low annealing temperature, compared to the PZT53/47 films. The perovskite phase of PZT53/47 films was not fully developed at a low annealing temperature of 500 °C, while the Ti-rich PZT films were relatively more crystallized into perovskite phase at 500 °C. Therefore, the Ti-rich films showed larger remanent polarizations than PZT53/47 films at a low temperature ranging from 500 to 550 °C. However, as the annealing temperature increases, the PZT53/47 films were fully crystallized into perovskite phase and exhibited largest remanent polarization.

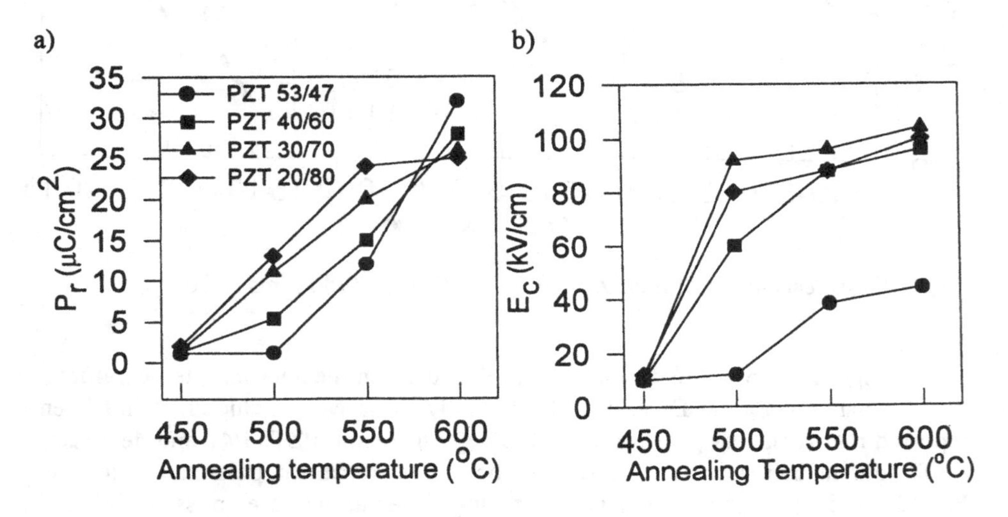

Fig. 3 a) remanent polarization (P_r) and b) coercive electric field (E_c) of PZT films as a function of annealing temperature.

Figure 3 b shows the dependence of coercive electric fields for various PZT films on the Ti content. The Ti content has strong influence on the coercive electric field. At both low and high annealing temperature regions, the coercive electric fields of Ti-rich PZT films were higher than that of PZT53/47, as observed in PZT bulks. High coercive electric fields for the Ti-rich PZT films might be attributed to the difficulty in domain switching in the tetragonal structure of the Ti-rich PZT films.

Current-voltage relationships were investigated at room temperature for the PZT films with various compositions annealed at 500 °C. Figure 4 shows the leakage current density of the PZT films as a function of applied voltage. Characteristic ferroelectric I-V curves were observed for the films, which exhibited good insulating properties even up tp an applied voltage of 8 V. The leakage current density of the PZT40/60 films was 10^{-7} A/cm^2 at an applied voltage of 10 V. It was observed that as the Ti content increased, the PZT films showed high leakage current density at the same applied voltage.

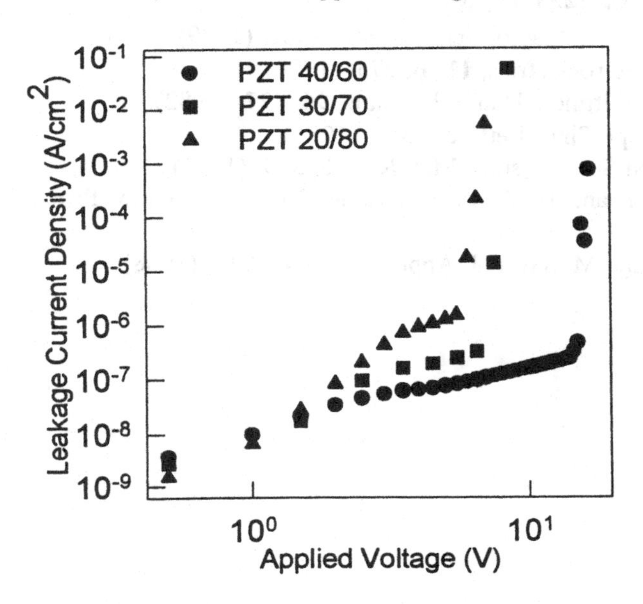

Fig. 4 Leakage current density of PZT films at 500 °C as a function of applied voltage.

CONCLUSIONS

Highly (111) oriented Ti-rich PZT thin films were successfully fabricated on $Pt/Ti/SiO_2/Si$ substrates at a low annealing temperature of 500 °C by a modified sol-gel technique. It was possible to obtain well-crystallized PZT films at 500 °C by modifying the precursor solution and directly annealing without pyrolysis step. The low temperature processing was assisted by taking advantage of the lattice matching of PZT (111) plane with $Pt/Ti/SiO_2/Si$ substrate, which reduces the activation energy for perovskite phase formation. The PZT films annealed at 500 °C exhibited excellent ferroelectric properties at an applied voltage of 5V. The remanent polarization and coercive electric field of the PZT films processed at 500 °C were found to increase as the Ti content increased. The leakage current density of the Ti-rich films was typically lower than 10^{-7} A/cm^2 at an applied voltage of 5 V.

REFERENCES

1. J. F. Scott, C. A. Araujo, H. B. Meadows, L. D. McMillan, and A. Schawabkeh, J. Appl., Phys., **66**, 1444, (1989).
2. J. F. Scott and C. A. Araujo, Science, **246**, 1400, (1989).
3. D. Bondurant, Ferroelectrics, **112B**, 273, (1990).
4. C. Y. Ting, M. Wittmer, Thin Solid Films, **96**, 327, (1982).
5. M. Wittmer, Appl. Phys. Lett., **37**, 540, (1980).
6. C. K. Kwok and S. B. Desu, J. Mat. Res., **8**, 339, (1993).
7. T. Tani, Z. Xu, and D. A. Payne, Mater. Res. Soc. Symp. Proc., **310**, 269, (1993).
8. G. Yi, Z. Wu, and M. Sayer, J. Appl. Phys., **64**, 2717, (1988).

TRIOL SOL-GEL ROUTE FOR PREPARING PZT THIN FILMS

S.J. Milne[1], R. Kurchania[1], J.D. Kennedy[2], M. Naksata[1], S. Arscott,[3] Duranee Kaewchinda,[1] Nimit Sriprang[1,2], Robert E. Miles[3], [1]Deptartment of Materials, [2]School of Chemistry, [3]School of Electrical and Electronic Engineering, University of Leeds, Leeds, LS2 9JT, UK.

ABSTRACT

Following on from our previous work on a diol sol-gel route for the fabrication of Pb $Zr_{0.53}Ti_{0.47}O_3$ (PZT) films in the thickness range 0.25-10 μm, we now report a related sol-gel route based on a triol, tris(hydroxymethyl)ethane. Good quality 0.4 μm films were prepared from a single application of the coating sol on platinized silicon substrates. Phase development and dielectric and ferroelectric properties are reported and compared with results for the diol sol-gel route. Data are also presented for triol sols deposited on platinized GaAs substrates. Despite the potential problems associated with Ga and As outdiffusion, ferroelectric PZT films with a remanent polarisation of 24 $\mu C\ cm^{-2}$ and coercive field of 32 $kV\ cm^{-1}$ could be prepared using rapid thermal annealing techniques.

INTRODUCTION

A number of sol- gel routes have been reported for the fabrication of thin films of lead zirconate titanate (PZT). Most of the extensive literature on sol-gel PZT thin films relates to systems employing metal alkoxides and carboxylates in a methoxyethanol (MOE) solvent, for which controlled hydrolysis and condensation reactions ultimately lead to the formation of a polymeric precursor sol[1]. The sols are coated onto a substrate, and after *in situ* gellation are thermally decomposed to form a crystalline ferroelectric film, the final heat treatment is usually carried out between 550°C and 700°C. More recently it has been shown that the limiting

To the extent authorized under the laws of the United States of America, all copyright interests in this publication are the property of The American Ceramic Society. Any duplication, reproduction, or republication of this publication or any part thereof, without the express written consent of The American Ceramic Society or fee paid to the Copyright Clearance Center, is prohibited.

crack-free thickness of single layer sol-gel derived PZT films can be increased, from an original value of 0.1 μm (on platinized silicon substrates), to 0.3μm by preparing sols using acetic acid in place of MOE [2,3]. For both types of sol, thicker films may be built up by employing repeated deposition / pre-thermolysis heat treatments, but the increased layer thickness of the acetic acid route makes it more attractive for fabricating multilayer films that may be used in pyroelectric and piezoelectric applications, for which films in excess of 1 μm are required.

In this laboratory we have explored the film forming capabilities of dihydric alcohols in conjunction with zirconium and titanium isopropoxides which have been modified with the complexing agent acetylacetone. The bidentate acetylacetonate (acac) ligand inhibits hydrolysis reactions relative to unmodified alkoxides, and by employing a combination of a diol solvent and stabilised metal propoxides we have successfully developed a sol-gel system which is capable of producing single layer PZT films with a thickness of 0.5-1 μm [4,5,6,7]. The diol is thought to exchange for isopropoxide ligands and eventually link together titanium and zirconium centers to produce a polymeric framework, possibly with dissolved lead acetate residing within the polymer matrix [8].

We have contiued to examine the potential of polyfunctional alcohols for the synthesis of PZT sols, and now after a systematic study of several potential systems, we report a new sol-gel route based on a triol, tris(hydroxymethyl)ethane, abbreviated THOME. Here we provide examples of the basic film forming capabilities of the new system for sols deposited on platinized silicon and platinized GaAs substrates.

EXPERIMENTAL

The starting reagents for sol synthesis were lead(II) acetate trihydrate {[$Pb(OOCCH_3)_2(H_2O)_3$], Alfa}, titanium(IV) diisopropoxide bisacetylacetonate {[$Ti(O^{iso}Pr)_2(CH_3COCHCOCH_3)_2$], Aldrich, abbreviated TIAA}, zirconium(IV) n-propoxide {[$Zr(O^nPr)_4$], Aldrich, 70 % w/w in 1-propanol}, 2,4-pentanedione {acetylacetone (Hacac), $CH_3COCH_2COCH_3$, Aldrich}, and 1,1,1-tris(hydroxymethyl)ethane {$CH_3C(CH_2OH)_3$, Aldrich, abbreviated THOME}. The details of the sol synthesis procedure are reported elsewhere [9] The viscous stock solution was stable for two to three days under ambient conditions, after which time a solid started to precipitate. However, dilution with methoxyethanol (MOE), to 0.65 - 0.70 M gave a more stable solution, and this was used for spin-coating onto Pt/Ti/SiO_2/Si or Pt/Ti/GaAs/Si_3N_4/SiO2/Si substrates of dimensions *ca.* 1cm x 1cm. The resulting gel films on silicon substrates were heat-treated in various ways. Low temperature (≤ 550°C) crystallisation studies were conducted by placing samples on a metal bock set on a laboratory hot- plate. A hole was drilled into the block in order to locate a thermocouple connected to a digital voltmeter.

This arrangement poduced a more stable heat source than simply placing the substrates on the conventional laboratory hot-plate. The surface temperature of a 'blank' platinized substrate, placed on the metal block was measured using a contact thermocouple; the blank was then replaced by a coated sample which was heated for 20 mins. Other samples were pre-treated at 350 °C and transferred to a tube furnace set at 500°C-600°C. Gel coatings on GaAs were processed using rapid thermal annealing, RTA, at 650°C for 1s, using a heating rate of 250 °C/s and a cooling rate of 70°C/s.

Phase analysis was carried out using X-ray diffraction (Phillips PW1820). The polarisation-electric field response (P-E) was determined with a Radiant Technologies RT66A ferroelectric tester at an applied frequency of 60 Hz for a triangular waveform of 300 kV cm^{-1}.

RESULTS AND DISCUSSION

In contrast to the previously developed diol sols, the properties of the new THOME sols are such that due to loss of MOE and other volatiles, the sol coatings gel *in situ* during the room-temperature spin coating phase of the process. This also occurs in the MOE and acetic acid sol-gel systems and makes it easier to produce physically uniform films. Moreover the chemical processing steps involved in the synthesis of THOME sols are more straightforward than for diol sols as the reaction was carried out in one step with no distillation of volatile by-products.

Preliminary NMR investigations indicate that the THOME species undergo ligand exchange reactions with propoxy and acac groups on the Zr and Ti starting reagents, eventually producing a sol consisting of bound acetate, acac and THOME in which, at this stage, we propose that gelation proceeds by the growth of THOME linked oligomers [9]. The lead component is probably retained initially as lead acetate in cavities within the THOME-linked Zr/Ti matrix.

Reactions between THOME and individual Ti and Zr alkoxides have previously been studied by other workers [10]; several interesting polynuclear compounds were isolated and their structures elucidated by single crystal X ray diffraction. However attempts at making thin films were unsuccessful which was attributed to the instability of the sol coatings toward atmospheric moisture. As far as we are aware there are no reports of using THOME in the more complex ternary PZT system to generate practicable ceramic films. Given the problems faced using Ti/Zr tetraisopropoxides it seems that the key to our success may lie in using acac as an alkoxide modifier to increase the resistance of the alkoxide toward hydrolysis.

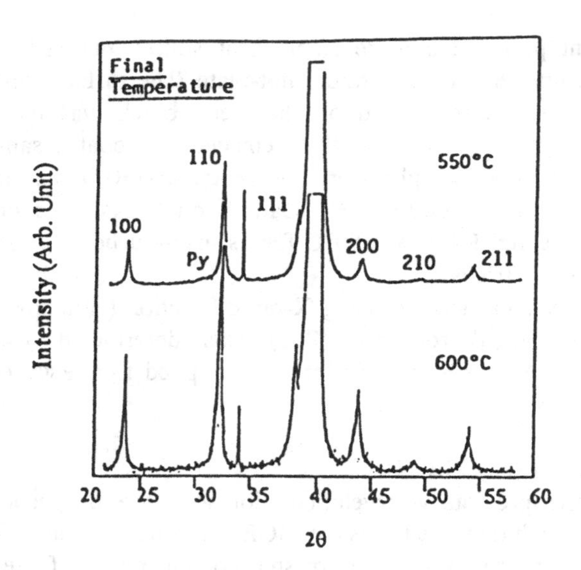

Fig 1. X-ray diffraction data for films fired at 550°C and 600 °C for 15 mins in a tube furnace (platinized silicon substrates)

Crystallisation of our gel coatings to predominantly perovskite PZT occured after heating in a tube furnace at 550 °C for 30 mins Fig 1, but heating at 600 °C was required to eliminate a faint extra peak at *ca.* 29.5 °2θ which is usually assigned to a pyrochlore or fluorite-type intermediate phase.

The thickness of a single layer film deposited from a 0.7M sol was 0.4 μm; an SEM examination of the surface microstructure of a sample made from sols containing 15 % lead acetate, and fired at 600 °C, implied a grain size of ca. 0.1 μm, but TEM investigations would be required to accurately evaluate the details of this fine grained microstructure.

In terms of polarisation-electric field response, the 550 °C sample exibited a broad and rather 'lossy' hysteresis loop. This is consistent with the prescence of the non-ferroelectric pyrochlore phase identified in the XRD pattern, and a very small grain size, <0.1 μm. An improved ferroelectric response was obtained for the sample fired at 600 °C, giving rise to a remanent polarisation value of 24 $\mu C\ cm^{-2}$ and a coercive field of 50 $kV\ cm^{-1}$, Fig 2.

When decomposition and crystallisation were carried out soley on a hot-plate, for which the measured substrate temperature was 540 ° C, the films displayed P-E

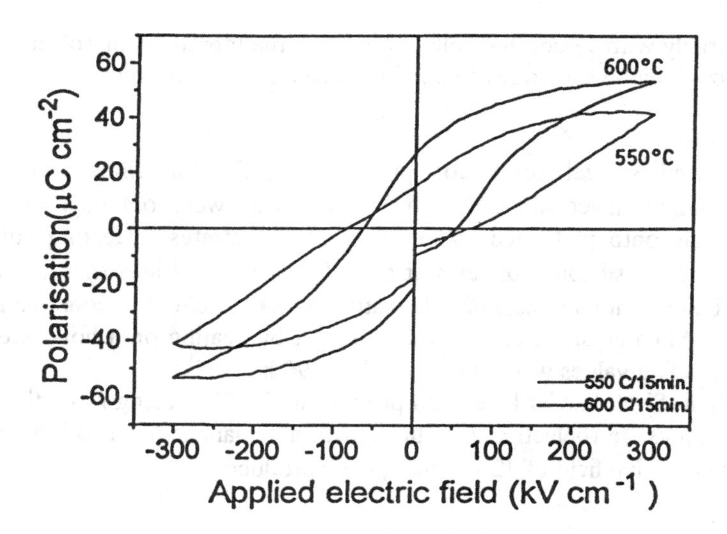

Fig 2. Polarisation-electric field data for 0.4 μm PZT films fired at 550°C and 600 °C for 15 mins.

hysteresis loops similar in general form to the loop obtained for the 600°C furnace-fired sample shown in Fig 2, but with a remanent polarisation of 36 μC cm^{-2} and a rather high co-ercive field, 90 kV cm^{-1}. We are continuing to investigate crystallisation behaviour under these conditions and will report the results in due course.

We proceeded to use the new sols in conjunction with rapid thermal annealing (RTA) in a project designed to fabricate PZT films on GaAs substrates. This configuration is of interest for microwave applications such as bulk acoustic wave resonators for direct integration into monolithic microwave integrated circuits. For sol-gel processing on GaAs substrates, it is essential to minimise the total thermal input so as to avoid severe problems of As and Ga out-diffusion and consequent PZT film degradation. It is also helpful to employ a diffusion barrier, in this case Si_3N_4 was used. Various RTA time-temperature combinations have been examined, and a narrow 'process window' has been identified enabling PZT films to be produced which display favourable ferroelectric properties. For example the P-E loop of a 0.4 μm PZT film produced by depositing a triol sol on Pt/Ti/Si_3N_4/GaAs followed by heating in the RTA at 650 °C for a dwell time of 1s is shown in Fig 3. The Pr and Ec values were 24 μC cm^{-2} and 32 kV cm^{-1} respectively, which compare favourably with values generally reported in the literature for sol-gel PZT films on platinized silicon substrates heated in the conventional manner.

CONCLUSIONS

A THOME based sol-gel route for processing PZT thin films has been demonstrated; single -layer films, 0.4 μm in thickness were obtained by spin coating 0.7M sols onto platinized silicon or GaAs substrates. Heating sol-gel coatings on platinized silicon substrates at 600 °C for 15 mins in a tube furnace, resulted in PZT films with a remanent polarisation of 24 μC cm^{-1} and coercive field of 50 kV cm^{-1}. When crystallisation was carried out by heating on a hot-plate at 540 °C corresponding values were 36 μC cm^{-1} and 90 kV cm^{-1}.

By employing a Si_3N_4 barrier layer, and processing the films using rapid thermal annealing techniques, ferroelectric PZT films with a remanent polarisation of 24 μC cm^{-1}, and a co-ercive field of 32 kV cm^{-1}, were produced.

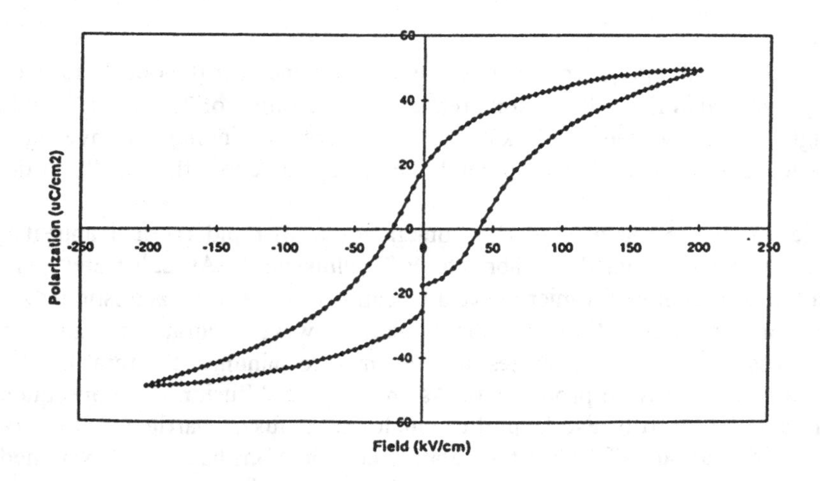

Fig 3. Polarisation-electric field data for 0.4 μm PZT film on $Pt/Ti/Si_3N_4/GaAs$ processed using rapid thermal annealing: heating rate, 250 °C/s; dwell at 650 °C for 1s; cool, 70 °C/s.

REFERENCES

[1]K.B. Budd, S.K. Dey and D.A. Payne, 'Sol-gel processing of $PbTiO_3$, $PbZrO_3$, PZT and PLZT thin films'. *British Ceramic Proceedings*, **36**, 107 -121 (1985)

[2]G. Yi, Z. Wu and M. Sayer, 'Preparation of PZT films by sol-gel processing' *Journal of Applied Physics*, **64**, 2717-23 (1988).

[3]R.A. Assink and R.W. Schwartz, '1H and ^{13} C NMR Investigations of PZT Thin Film Precursor Solutions,' *Chemistry of Materials*, **5** [4] 511-517 (1993)

[4]N. J. Phillips and S. J. Milne, 'Diol based sol-gel system for the production of thin films of $PbTiO_3$' *Journal of Materials Chemistry*, **1** 893-894 (1991).

[5]Y. L. Tu, M. L. Calzada, N. J. Phillips and S. J. Milne, 'Synthesis and electrical characterisation of thin films of PT and PZT made from a diol based sol-gel route,' *Journal of the American Ceramic Society*, **79** 441-448 (1996)

[6]Y.L. Tu, and S.J. Milne, 'Characterisation of single-layer PZT films prepared from an air-stable sol-gel route,' *Journal of Materials Research*, **10** 3222-3231 (1995)

[7]Y. L. Tu and S. J. Milne, 'Processing and characterisation of PZT films up to 10 μm thick,' *Journal of Materials Research*, **11**, 2556-2564 (1996)

[8]N. J. Phillips, N. J. Ali, S.J. Milne and JD Kennedy, 'A preliminary NMR investigation of the titanium diisopropoxide bis 2,4 pentanedionate-propanediol sol-gel system,' *Journal of Materials Science*, **13** 1535-1537 (1994)

[9] N Sriprang, D. Kaewchinda, J.D. Kennedy and S.J. Milne. 'Triol based sol-gel route for preparing PZT thin films,' *Journal of the American Ceramic Society* (in review)

[10]T. Boyle, R Schwartz, R. J. Doedens and J W Ziller, 'Synthesis and Structure of Novel Group IV Tridendate Alkoxide Complexes and Ceramic Films Derived Therefrom,' *Inorganic Chemistry*, **34** 1110-1120 (1995)

ACKNOWLEDGEMENTS

This work was funded in part by the UK Engineering and Physical Sciences Research Council; we also wish to thank the Royal Thai Government for providing student scholarships.

PHASE TRANSFORMATIONS AND ORIENTATION DEVELOPMENT IN MMAP DERIVED PZT (40/60) THIN FILMS

Pradeep P. Phulé and Yimin Liu
Department of Materials Science & Engineering
848 Benedum Hall, University of Pittsburgh
Pittsburgh, PA 15261

ABSTRACT

A sol-gel process, based on the molecularly modified alkoxide precursors (MMAP) approach, was used for the preparation of dense, fine-grained (~ 100 nm), single phase perovskite thin films of PZT (40/60). The sequence and kinetics of the transformation of amorphous thin layers into perovskite PZT were investigated using X-ray diffraction (XRD) and transmission electron microscopy (TEM). Based on the results, a model concerning texture selection and phase formation sequence in sol-gel derived PZT thin films is proposed. It was found that by including or excluding a 400°C intermediate heat treatment step, crystallized PZT thin films exhibited a strong (111) or (100) texture, respectively. The observations made at different stages of thermal processing indicated that the films stayed amorphous up to 400°C/30min. during pyrolysis, while the highly textured, single phase perovskite films were obtained upon 700°C treatment for times as short as 1 minute. The (111) texture development appeared to be controlled by nucleation of a perovskite phase, assisted by a solid phase epitaxial (SPE) effect, and the (100) and texture appeared to be growth controlled. This model suggests that similar to the hydrothermal and other chemical routes, conversion of the amorphous films into the perovskite phase, without the formation of pyrochlore or similar other intermediate phases, is possible.

INTRODUCTION

Owing to their unique ferroelectric and other properties a high level of scientific and technological interest currently exists in the area of thin films of ferroelectric materials [1]. In particular, the field of lead zirconate titanate $Pb(Zr_xTi_{1-x})O_3$ (PZT) thin films has witnessed extensive research and development efforts with a view to develop such devices as non-volatile ferroelectric random access

To the extent authorized under the laws of the United States of America, all copyright interests in this publication are the property of The American Ceramic Society. Any duplication, reproduction, or republication of this publication or any part thereof, without the express written consent of The American Ceramic Society or fee paid to the Copyright Clearance Center, is prohibited.

memories (nvFRAMs), dynamic random access memories (DRAMs), micro-sensors and micro-actuators[2].

In addition to the requirement on film/substrate compatibility, the "thermal budget" for ferroelectric thin film processing must be reduced to protect underlying layers during fabrication, meanwhile the quality of the films and their ferroelectric properties must not be compromised. Therefore, an understanding of the phase formation and microstructure development during the fabrication of ferroelectric thin films is valuable in realizing these goals. Many factors with respect to process conditions e.g. heat treatment, composition, stoichiometry, and substrate, are known to influence the pathways to ferroelectric perovskite phase formation and microstructure.

In thin film PZT processing it is generally believed that the final perovskite structure is achieved via an intermediate pyrochlore or similar intermediate phase. In this paper, we report studies on the phase evolution at different stages of thermal processing for the sol-gel derived PZT thin films. These studies demonstrate the possibility of a direct, amorphous → perovskite transformation, without the pyrochlore-like phases as *necessary* intermediates. Another important microstructural feature of ferroelectric PZT thin films is the preferred orientation of grains (i.e., texture) and in this paper we also have made an effort to establish the nexus between the sequence of phase transformation and orientation development in MMAP derived PZT thin films.

EXPERIMENTAL

Composition and Sol Preparation

The composition chosen was $Pb(Zr_{0.4}Ti_{0.6})O_3$ (PZT 40/60). This composition lies in the tetragonal region of the PZT phase diagram. The ferroelectric hysteresis loop for this composition is expected to be square, which is beneficial for the memory applications. For sol preparation, titanium isopropoxide ($Ti[OCH(CH_3)_2]_4$, Aldrich) was added to isopropanol (Aldrich). Glacial acetic acid (CH_3COOH, Fisher Scientific) was slowly added to the sol with an (acetic acid)/(B-site cations) molar ratio of 16. The sol was stirred for about 1 hour. Separately, *hydrated* lead acetate ($Pb(CH_3CO_2)_2{\cdot}3H_2O$, Aldrich, with Pb/(Ti+Zr) molar ratio of 1.1, or 10% excess lead) was dissolved in 2.0 ml. glacial acetic acid and stirred. After stirring for 1 hour, zirconium acetylacetonate ($Zr[CH_3COCHCOCH_3]_4$, Aldrich) was added to the modified Ti-precursor sol with a Zr/Ti molar ratio of 2/3. After another hour of stirring, the precursor sols were combined and mixed for 1 hour. These steps were performed at room temperature in a glove box (Vacuum Atmospheres HE-493) containing dry argon atmosphere.

Spin-Coating and Heat Treatment

Ramtron Corporation supplied most substrates used in this study. These consisted of a ~ 150 nm, Pt (111) oriented electrode, and a ~ 20 nm Ti adhesion layer. These layers were grown on amorphous SiO_x (~ 500 nm). The Si crystal was (100) oriented and ~ 350 μm thick.

The sol was applied through a syringe with a 0.22 μm filter (Micro Separations, Inc.). After completely covering the substrate (~ 3 ml. of sol), a thin film was spin coated onto the substrates using a photo-resist spinner (Headway Research Inc. 1-EC101D-R485) by spinning the sol for 10 seconds at a speed of 4000 rpm. Deposited films were placed on a hot plate at 300°C for 2 minutes for removal of the organic solvent. This procedure was repeated three times before further heat treatment. The film obtained at this stage is referred to as one layer. The 300°C heat treatment was performed three times per layer. Films were crystallized at a final temperature of 500-700°C for a duration of 1 to 30 minutes, by directly inserting the samples into a preheated furnace in air or oxygen atmosphere. The temperature near the sample was monitored using a separate thermocouple. An intermediate heat treatment at 400°C for 2-30 minutes in a preheated furnace was employed for selected samples. Whenever necessary, more layers were deposited onto previously crystallized layers and heat treated as needed.

X-ray Diffraction (XRD) and Scanning and Transmission Electron Microscopy Analysis

Films obtained during different stages of processing were characterized by XRD (Phillips XPERT). Two different diffraction geometries (Bragg-Brentano and glancing angle) were used to study the thin film crystallization and microstructure. A tube voltage of 30 kV and a current of 20 mA were used for Bragg-Brentano geometry. For glancing angle geometry, a tube voltage of 40 kV and a current of 30 mA were used. CuKα radiation was used in both cases.

Microstructural features such as surface morphology, grain size, and thickness, were observed using SEM and TEM. For some SEM samples a thin layer of Au was deposited so as to reduce charging of the samples. The SEM (Phillips Model XL30) was operated at 5-25 kV. Electron transparent TEM cross-sectional samples were prepared using mechanical polishing (Buehler Minimet Polisher), followed by dimpling (Gatan Model 656 Dimpler), and ion milling (Gatan Model 600 dual ion mill, Gatan PIPS™ Model 691 precision ion polishing system). A beam current of 0.5 mA and a gun voltage of 3-5 kV were used during ion milling. The TEM (JEOL 200CX) and STEM (JEOL 2000FX) were operated at 200 kV.

RESULTS AND DISCUSSION

XRD analysis was conducted at different stages of thermal treatment, namely, after 300°C hot plate drying, 400°C pyrolysis at different times (2, 5, 10 and 30 minutes), and after final crystallization at 700°C. Before the final crystallization, all XRD patterns obtained were essentially the same, the only peaks observed were those related to the substrate. There was no indication of either pyrochlore or perovskite formation, films at these stages were still amorphous, within the detection limit of XRD. Glancing angle XRD was also conducted on these films and pyrochlore or perovskite phase formation was not observed.

We also examined the crystallization of PZT films for crystallization times of 1, 2, 5, and 10 minutes. In each case, the films were heat treated by directly inserting the samples into a pre-heated furnace maintained at 700°C. The samples were air quenched following this heat treatment. The XRD pattern of a film after heat treated at 700°C for 1 minute showed that even after such a short heat treatment, a sharp peak corresponding to perovskite PZT (111) developed. No other PZT related phases could be detected using XRD.

Figure 1 shows a SEM cross-sectional view of a 3 layer PZT 40/60 thin film, with an overall thickness of ~ 300 nm. Figure 1 also shows a SEM planar view of a two-layer PZT film crystallized at 700°C for 30 min. As can be seen from this micrograph, the crystallized films had a dense, fine-grained structure with grain size of about 100 nm. No significant differences in the surface morphology were observed for PZT heat treated at 500, 600 or 700°C.

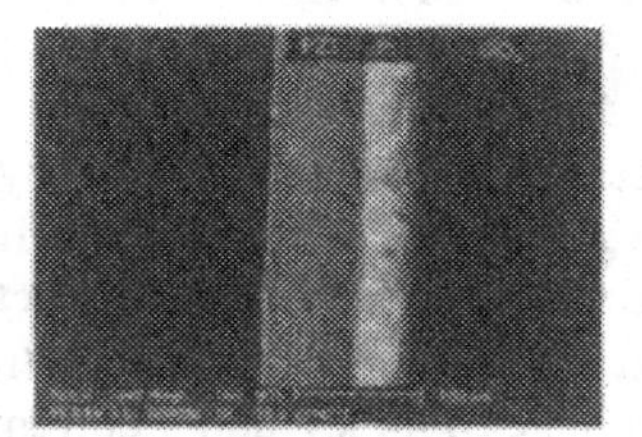

Figure 1. Scanning electron micrographs of a cross-sectional and plan views of a typical MMAP derived PZT (40/60) thin films.

A bright field image of a PZT thin film heat treated at 400 ° C for 30 minutes is shown in Figure 2a. Selected area diffraction patterns and the images of films treated under such conditions did not reveal any microcrystalline pyrochlore or other phases. Figure 2b shows a dark field TEM image of a PZT thin film crystallized at 700 ° C for 1 minute (film thickness ~ 100 nm). Electron diffraction as well as XRD on such films confirmed the presence of perovskite PZT.

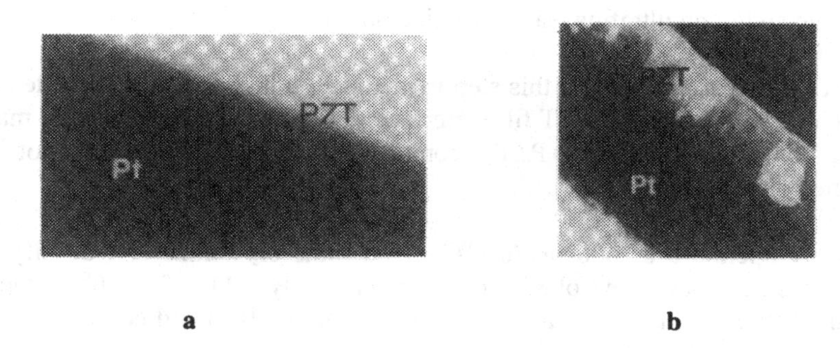

Figure 2. TEM images of a PZT thin film heat-treated at (a) 400 C, 30 min. and (b) 700 C, 1 minute. Thickness of the film is ~ 100 nm.

Notice that even after times as short as one minute, the PZT thin films heat treated at 700 °C had undergone crystallization and exhibited columnar grains. Pole figure analysis on such films revealed a (111) fiber texture[3].

The orientation development in MMAP derived thin films of PZT was also found to be closely linked to the heat treatment schedule employed. For fixed final crystallization at 700°C and 30 min., films processed with or without an intermediate heat treatment step at 400°C/30 min. developed two distinct textures. On inclusion of a 400 °C step, PZT 40/60 films with a strong (111) texture were observed. When the 400 °C step was excluded, the film exhibited a strong (100) texture. The PZT (111) preferred orientation was also obtained for films crystallized at 550, 600 or 700 °C, following a 400 °C/10 min. heat treatment step. The (111) or (100) texture developed under these different processing conditions was further confirmed by pole figure analysis. These observations were reproduced in at least 25 independently prepared samples[4].

Model for Texture Evolution and the Sequence of Phase Formation

Based on our results, we propose the following model regarding crystallization process and phase transformation sequence in MMAP derived 40/60 PZT thin films.

Step 1 : Temperature 300 °C: An amorphous gel network consisting of Zr-O-Zr and Ti-O-Ti linkages evolves. At this stage, the MMAP derived films contain lead as lead acetate and/or an ultrafine lead oxide like phase.

Step 2: Temperature 400° C: If this step is included, a layer of (111) oriented PZT seed nuclei evolves at the Pt/PZT film interface. It is possible that the nuclei may be Ti- rich compared to overall PZT composition, however, this has not been determined directly.

Step 3: Temperature of 600 or 700 °C: The films crystallize and densify. The orientation is nucleation controlled and is predominantly (111), if a 400° C step was included. Otherwise, the orientation is predominantly (100) and controlled by the kinetics of the growth process.

Based on our experimental results and in conjunction with the evidences found in the literature, we believe that in the sol-gel process used in this work, formation of perovskite PZT via a pyrochlore phase formation is not favored. There are several reasons underlying this: (1). The nucleation energy for perovskite crystallization is lower for titanium rich PZT (e.g. 40/60) compositions; also additional driving force for perovskite nucleation is provided by the epitaxial match with (111) Pt. (2). Excess lead is used to suppress the formation of lead deficient pyrochlore phase. (3) Using the direct insertion method, heating rate is high, and the temperature range for pyrochlore formation is bypassed quickly. (4). In MMAP approach, the Zr and Ti cations are incorporated in the gel-network and hence can be considered mixed at a truly molecular level. On the other hand, lead, introduced in our process as hydrated lead acetate, remains outside the metal-oxygen-metal network. In summary, based on our XRD and electron microscopy observations, other reports in the prior literature we believe that MMAP derived PZT 60/40 thin films can rapidly crystallize into oriented, perovskite PZT without forming pyrochlore-like phases as reactive intermediates.

ACKNOWLEDGMENTS

We are grateful to Dr. Lee Kammerdiner for providing us with the substrates used in this study. One of the authors (PPP) acknowledges support from the School of Engineering at the University of Pittsburgh through a William Kepler Whiteford Faculty Fellowship and financial support through grants from the Division of Materials Research, National Science Foundation (Grant Numbers: DMR 94-15567 and DMR 95-04015).

REFERENCES

[1] S .K. Dey, " Sol-Gel Science and PE-MOCVD of Dielectric Perovskite Films in Ferroelectric Thin Films: Synthesis and Basic Properties", in *Ferroelectrics and Related Phenomena* Eds. C. Paz de Araujo, J.F. Scott, and G.W. Taylor, 10, 329-389, Gordon & Breach publishers, Netherlands (1996).

[2] See Special Issues of the MRS Bulletin, 21 [6, 7] (1996).

[3] Y. Liu and P. P. Phulé, J. Am. Cer. Soc, 79 [2] 495-98 (1996).

[4] P. P. Phulé and Y. Liu, J.Am.Ceram Soc. 80 [9] 2410-12 (1997).

PREPARATION OF PZT THIN FILMS ON HASTELLOY SUBSTRATE USING SOL-GEL PROCESSING

W. Yu, B. A. Chin, Z. Chen, Auburn University, Auburn, AL 36830

ABSTRACT

Ferroelectric PZT thin films have been successfully coated on Si substrates using sol-gel technology for sensing applications. However little work pertinent to PZT film coating on metallic substrates has been reported. Hastelloy has good high temperature oxidation resistance, good electrical conductivity, and high mechanical strength. Due to these excellent properties, Hastelloy was chosen as a substrate material for sol-gel derived PZT film for sensing applications. However, the PZT precursor reacted with Hastelloy, inhibiting the formation of PZT perovskite structure. A thin intermediate layer was therefore introduced as a reaction barrier layer to reduce the extent of Hastelloy substrate-precursor solution reaction. Results showed that the reaction barrier layer significantly reduced the reaction. Thus the PZT films with the perovskite structure, were successfully formed on the Hastelloy substrate. X-ray diffraction analysis was conducted to study the effect of using reaction barrier layers on the film structure.

INTRODUCTION

Piezoelectric PZT in film form has been applied to positioners, sensors and actuators[1,2,3]. PZT thin film devices that require much lower driving voltages, and a greater precision of control thus can be achieved as compared to devices fabricated from traditional sheet PZT. The successful deposition of PZT thin films on various substrates is one of the essential technologies, which will make these applications possible.

There are several ways to prepare PZT films. PZT films have been prepared by screen-printing, electron beam evaporation[4], RF sputtering[5], ion beam deposition[6] and hydrothermal[7] growth. Most of these techniques require multi-component material from source materials. A complicated set of parameters including substrate temperature, chamber pressure, gas composition,

To the extent authorized under the laws of the United States of America, all copyright interests in this publication are the property of The American Ceramic Society. Any duplication, reproduction, or republication of this publication or any part thereof, without the express written consent of The American Ceramic Society or fee paid to the Copyright Clearance Center, is prohibited.

sputtering power, and the composition of the target, have to be optimized. Sol-gel technology arose as an alternate method of fabricating high quality ceramics and glass as well as thin films. Sol-gel processing has many advantages over other methods. The method enhances purity and homogeneity of the film composition, enables the reduction of processing temperatures, promotes easy formation of thin layers, and reduces processing expense.

The most common sol-gel processed metal oxides, are derived from alkoxide solutions. A mixture of alkoxides, solvents and additions is called a sol. The sol undergoes hydrolysis and condensation to form a cross-linked structure called a gel, which is capable of immobilizing the remaining solvents. Upon heating, additions and solvents evaporate from the gel, and the cross-linked network undergoes pyrolysis and oxidation, leaving a porous metal oxide film.

Budd et al were the first to report work on depositing PZT thin film using sol-gel techniques in 1985.[8] Since then, much work has been devoted to the preparation of PZT films on various substrates. Studies of the deposition of PZT films on Si wafer with various intermediate layers have been conducted [1,9]. Experiments on the deposition of PZT thin film on GaAs substrates using the sol-gel process for random access memories applications have been reported[10]. Studies of the deposition of PZT film on glass and aluminum substrate with an Al_2O_3 intermediate layer were done by Yi et al[11,12] using the sol-gel process. Investigations on sol-gel processing for the deposition of PZT films on an Al_2O_3 substrate and on a MgO substrate were performed by Schonecker et al[13] and Tuttle et al[14] respectively. In the same paper by Tuttle, deposition on Pt was discussed. However, there are very few publications regarding PZT thin film deposition using sol-gel process on metallic substrate other than on Pt. PZT thin film deposition on stainless steel was mentioned in one of Sayer's papers with no details.

Using sol-gel technology to coat PZT thin film on metallic substrate can make piezoelectric devices which require much lower actuation voltages, compared to devices fabricated from traditional sheet PZT. The direct deposition of PZT on a metal foil makes use of the substrate as a bottom electrode. This eliminates the need of making an additional bottom electrode, as in the case of using a non-conductive ceramic or a semi-conductive substrate, thus simplifying the procedure of fabricating a miniature sensor. However, using the sol-gel method to coat PZT precursors on metallic foil has resulted some difficulties in the past. The combination of environmental instability of the metallic foil and extremely chemically active species in PZT precursors can inhibit the formation of PZT perovskite structure. In this paper, we investigate the possibility of forming PTZ film on a Hastelloy substrate.

EXPERIMENTAL PROCEDURES

Water based PZT precursor similar to that used by Yi's[13] was used to prepare PZT (Zr/Ti=53:47) film. The precursor is a mixture of titanium (IV) isopropoxide (99.999%), zirconium (IV) propoxide (70% 1-propal), glacial acetic acid, lead (II) acetate trihydrate (99+%), distilled water, lactic acid, and glycerol. An Inverse Mix Order (IMO) method was adopted[15]. Many metallic foils have been screened for substrate candidates, such as Al, Cu, Ti, Nb, Hastelloy. Platinum was not selected for this study due to its high cost. Hastelloy (Ni-Cr-Mo alloy) was used as the PZT substrate for its high environmental resistance. A Headway Photo Resistance spinner was employed to spin coating precursor onto the substrate. Spin coatings were typically formed at the speed range from 2000 to 4000 RPM for 30 seconds. Following spin coating, samples were heat-treated in a tube furnace in an atmosphere. X-ray diffraction (XRD) was used for structural characterization.

RESULTS AND DISCUSSION

PZT Precursor Direct Coating on Hastelloy

Figures 1(a) and (b) show the x-ray diffraction pattern for the as-received Hastelloy and heated Hastelloy substrates respectively. The Hastelloy substrate was heated to 400°C, held for 30 minutes, then rapidly heated to 700 °C and held for 3 hours. From the x-ray results no additional peaks were observed. This suggests Hastelloy's excellent oxidation resistance.

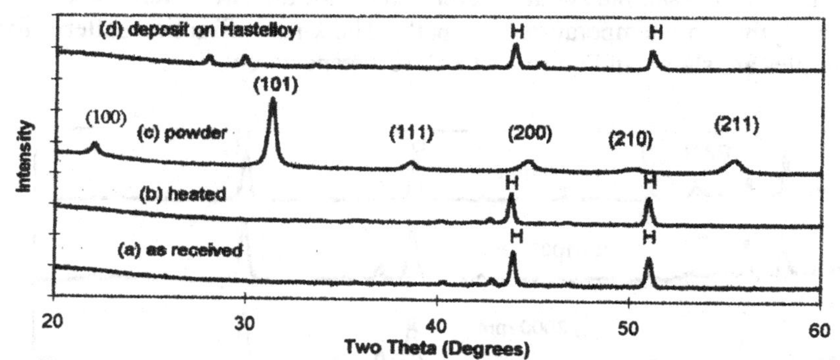

Fig. 1. XRD pattern (a) as received Hastelloy substrate, (b) heated Hastelloy substrate, (c) powder, (d) deposit on Hastelloy substrate

Figure 1(c) is the x-ray diffraction pattern from powdered PZT, which was obtained from the prepared precursor. The precursor was heated in a tube furnace to 400°C, held for 30 minutes, and subsequently was heated to 700 °C for 3 hours. All the peaks for PZT perovskite structure were identified. The PZT peaks do not overlap with any of the peaks from the Hastelloy substrate.

The as-received Hastelloy foil was cleaned using propanol prior to coating with PZT precursor. The coated sample was heat-treated in a tube furnace at 400°C for 30 minutes, following which the temperature was raised to 700 °C and held for 1 hour. Figure 1 (d) is the x-ray diffraction pattern of the coated substrate after heat treatment. In Figure 1 (d), two new peaks have appeared in addition to the Hastelloy peaks. The two additional peaks do not correspond to any peaks of the PZT perovskite structure. The additional peaks indicate that chemical reaction(s) have occurred between the PZT precursor and the Hastelloy substrate during the post heat treatment process. The elements in the precursor were consumed by the reaction to form additional phase(s). Therefore, it is highly probable that the chemical balance in the precursor was redistributed. As a result, no PZT perovskite phase was formed.

The type of reaction occurring has not been identified yet. However, it is interesting to understand how much of the elements can be lost during the coating and heat treatment procedures, so we can control or limit the reaction. If the reaction is limited only on the contacting interface between the Hastelloy and the precursor, a thicker coating may be able to help form the PZT structure. Several samples were prepared at different spin speeds (2000 rpm, 1000 rpm and 60 rpm) to achieve precursor coatings of different thickness.

Three coated samples with different precursor thickness were then heat-treated using the same temperature-time path. The x-ray diffraction patterns for these samples are shown in Figure 2(a), (b), (c) respectively.

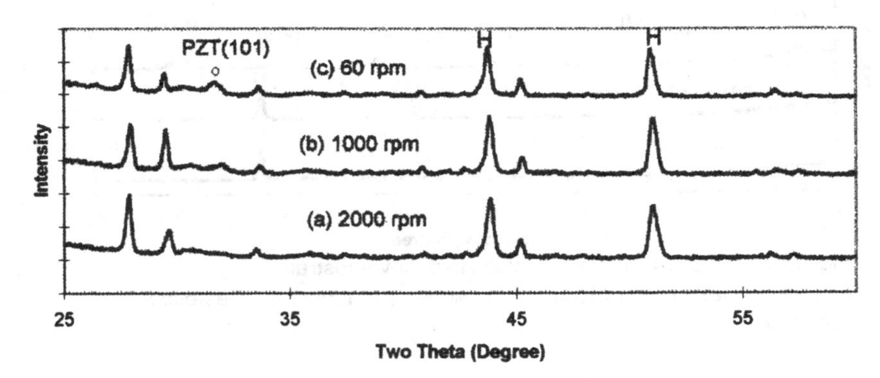

Fig. 2. Deposit on Hastelloy (a) 2000 rpm, (b) 1000 rpm, (c) 60 rpm

As we can see from the x-ray pattern, with increasing precursor coating thickness, the reaction between the substrate and the precursor persisted and no significant PZT peaks were observed. The only PZT peaks observed were from the sample which was coated at the lowest speed - 60 rpm. The sample coated at the spin speed of 60rpm shows the appearance of the strongest perovskite peaks (2θ about 31°). The results indicated that a certain amount of precursor needs to be sacrificed in order to form a PZT perovskite structure on the outer surface.

Oxidation Layer as a Diffusion Layer

The results above indicate that some chemical reaction(s) occurred between the Hastelloy substrate and the precursor. Generally speaking, Hastelloy foil has good acid resistance and environmental stability. However, it was suspected that an electrochemical reaction occurred during coating and the initial drying process. Metal passivation has often been used in industry as a way to avoid or reduce electrochemical reaction. Most metals near the anodic end of the galvanic series are active and serve as anodes in most electrolytic cells. However, if these metals are made passive, or cathodic, they will corrode at a much slower rate than normal. The easy way to produce passivation on the metallic surface is to expose it in an oxygen-rich environment. We can apply the same principle in coating a precursor onto the Hastelloy substrate. An as-received Hastelloy foil was pre-oxidized at 500 °C for 30 minutes to form a very thin yellowish layer. The film was then spin coated at 2000 rpm and then heat-treated at the same schedule. The resulting x-ray diffraction pattern is shown in Figure 3.

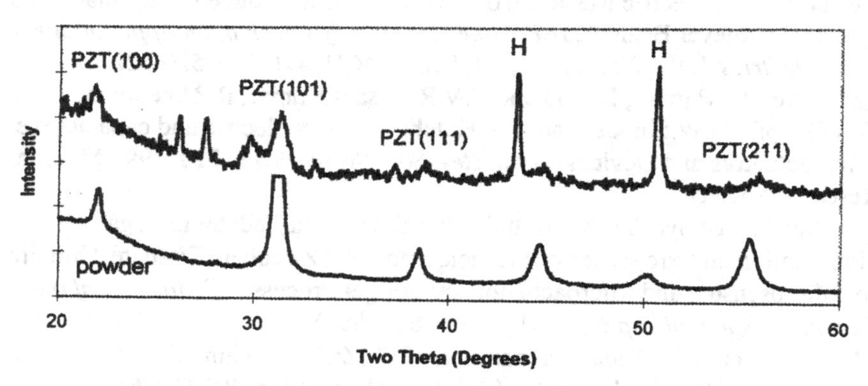

Fig. 3. Deposit on preoxidized Hastelloy substrate

All PZT perovskite peaks were identified, although the peaks from the substrate still dominated in intensity due to the high relative thickness of the substrate. Two unidentified additional peaks were observed in its XRD spectrum. However, comparing to the sample using a non-passivated Hastelloy foil with the same coating parameters, the reaction between the substrate and the precursor was significantly reduced. As a result, the PZT film was obtained. We conclude that the thin passive layer limits the chemical reaction between the Hastelloy substrate and the PZT precursor, thus the PZT perovskite phase can be formed on the Hastelloy substrate.

CONCLUSIONS

Some of the difficulties encountered during coating PZT precursor on Hastelloy have been recognized. Although Hastelloy generally has good environmental stability and good acid resistance, the PZT precursor is chemically very active and readily attacks the metallic elements in Hastelloy. However, by adding a passivation process, it is feasible to coat PZT film on a Hastelloy substrate using sol-gel process.

REFERENCES

[1] K.G. Brooks, D. Damjanovic, and N. Setter, Ph. Luginbuhl, G.A. Racine, N.F. DeRooij "Piezoelectric Response of PZT Thin Film Actuated Micromachined Silicon Cantilever Beams", *IEEE International Symposium on Applications of Ferroelectrics 1994*, Piscataway, NJ, USA, 94CH3416-5, p. 520-522.
[2] M. Sayer, D. Barrow, L. Zou and C.V.R.Vasant Kumar, R. Noteboom, D.A.Dnapik, D.W.Shindel and D.A.Hutchins, "Piezoelectric and capacitative Microactuators and Devices", *Mat. Res. Soc. Symp. Proc.* **310**. 1993 Materials Research Society
[3] Toshiyuki Tuchiya, Toshihiro Itoh, Gen Sasaki and Tadatomo Suga, "Preparation and Properties of Piezoelectric Lead Zirconate Titanate Thin Films for Microsensors and Microactuators by Sol-gel Processing", *Journal of the Ceramic Society of Japan*, **104** [3] 159-163 (1996)
[4] M. Oikawa and K. Toda, "Preparation of $Pb(Zr,Ti)O_3$ Thin Film by an Electron Beam Evaporation Technique", *Appl. Phys. Lett.* **29**, p. 491 (1976)
[5] A. Croteau and M. Sayer,"Growth and Characterization of $Pb(Zr,Ti)O_3$ in *Proceedings of the 6th IEEE International Symposium on Applications of Ferroelectrics*, June 8-11, 1986, Bethlehem, PA, edited by V.E.Wood (IEEE, New York, 1986), pp.606-609.

[6] R. N. Castellano and L. G. Feinstein, "Ion-beam Deposition of Thin Films of Ferroelectric Lead Zirconate Titanate (PZT)" *J. Appl. Phys.* **50** [6] 4406-4411 (1979)
[7] Takayuki Kikuchi, Takaaki Tsurumi, Yoki Ohba, "Bending Actuator Using Lead Zirconate Titanate Thin film Fabricated by Hydrothermal Method", *Jpn. J. Appl. Phys.* **31** part I [9B] 3090-3093, (Sept. 1992)
[8] K. D. Budd, S. K. Dey and D. A. Payne, "Sol-gel Processing of $PbTiO_3$, $PbZrO_3$, PZT and PLZT Thin Films", in *Brit. Cer. Soc. Proc.*, **36**, 1985, pp.107-121
[9] Charles D. E. Lakeman, Jeon-Florent Campion, Carlos T. A. Suchicital and David A. Payne, "An Investigation into the Factors Affecting the Sol-gel Processing of PZT Thin Layers", *1990 IEEE 7th International Symposium on Applications of Ferroelectrics*, p. 681-684.
[10] W. A. Geideman, S.Y. Wu, L.E. .Sanchez, B. P. Maderic, W. M. Liu, I. K. Naik and S.H. Watanabe, "PZT Thin Films for GaAs Ferroelectric RAM Applications", *90 IEEE 7th International Symposium on Applications of Ferroelectrics*, Champaign, IL, USA, p. 258-262.
[11] Guanghua Yi, Zheng Wu, and Michael Sayer, "Preparation of $Pb(Zr, Ti)O_3$ Thin Films by Sol Gel Processing: Electrical, Optical, and Electro-optic Properties", *J. Appl. Phys*, **64** [5] 2717-2724 (1988).
[12] Guanghau Yi, and M. Sayer, "Sol gel processing of thick PZT films", *Proceedings of the 1992 IEEE International Symposium on the Applications of Ferroelectrics*, p.289-292
[13] A. Schonecker, H.J. Gesemann, S. Merklein, W. Grond, K. Franke, M. Weihnacht, "PZT-film compositional development and physical properties", *1994 IEEE International Symposium on Applications of Ferroelectrics*, CH3416-5, p. 412-415.
[14] Bruce A. Tuttle, James A. Voigt, Terry J. Garino, David C. Goodnow, Robert W. Schwartz, Diana L. Lamppa, Thomas J. Headley, and Michael O. Eatough, "Chemically prepared $Pb(Zr,Ti)O_3$ thin film: the effects of orientation and stress", *1992 IEEE International Symposium on Applications of Ferroelectrics*, p. 344-348.
[15] B.A. Tuttle, J.A. Voigt, D. C. Goodnow, and D. L. Lamppa, "Highly Oriented, Chemically Prepared $Pb(Zr, Ti)O_3$ Thin Films." *Journal of the American Ceramic Society*, **76**[6] 1537-1544 (1993)

SYNTHESIS AND CHARACTERIZATION OF WET-CHEMICALLY DERIVED STRONTIUM BISMUTH TANTALATE (SBT) THIN FILMS

J.T. Dawley, R. Radspinner, B.J.J. Zelinski, D.B. Hilliard, K.A. Jackson, G. Teowee, and D.R. Uhlmann

Department of Materials Science and Engineering
University of Arizona, Tucson, AZ 85721

P.Y. Chu, B.M. Melnick, and R.E. Jones, Jr.

Advanced Materials Group, MRST
Motorola, Austin, TX 78721

ABSTRACT

Because of its fatigue resistance, large remanent polarization, and low coercive field, strontium bismuth tantalate (SBT) qualifies as a candidate for nonvolatile memory applications. It has been proposed that SBT's unique ferroelectric properties are a result of its layered perovskite structure, which controls the formation of defects that lead to fatigue, and creates a large spontaneous polarization due to the underbonding of bismuth atoms in the perovskite A site. Solution deposition techniques for SBT provide a fast and inexpensive means to evaluate a wide range of compositions and permit the control of composition and microstructure. Evidence in the literature on SBT synthesis indicates that the details of the synthesis route can have a major impact on film properties. In this work, the ferroelectric behavior of SBT films is characterized for a new alkoxide-acetate solution route. Emphasis is placed on establishing relationships between synthesis route, microstructure, and ferroelectric properties.

INTRODUCTION

Strontium bismuth tantalate ($SrBi_2Ta_2O_9$) belongs to a group of compounds known as layered perovskites, having the chemical formula $A_{m-1}Bi_2B_mO_{3m+3}$.[1] The structure of these compounds contains alternating perovskite-like layers of the composition $\{A_{m-1}B_mO_{3m-1}]^{2-}\}$ and rock salt structured $(Bi_2O_2)^{2+}]_\infty$ layers.[2, 3] In stoichiometric $SrBi_2Ta_2O_9$, one complete perovskite block of Ta-O octahedra and Sr^{2+} cations is sandwiched between two $Bi_2O_2^{2+}$ layers. Displacements of the Sr^{2+} and Bi^{3+} cations, along the a-axis of the perovskite layers, produce the anisotropic polarizability observed in SBT. In the bulk, the dielectric constant of SBT is 180. When cooled below the Curie temperature of 310-335 C, SBT has a spontaneous polarization of 14.4 $\mu C/cm^2$ and a piezoelectric strain coefficient (d_{31}) of 23 pC/N.[4-7]

Wet chemical or sol-gel techniques have been used to deposit a wide variety of thin film materials. Such techniques provide atomic-level control of film composition, as well as rapid and inexpensive techniques for evaluating compositional variations and the ability to obtain desired film

To the extent authorized under the laws of the United States of America, all copyright interests in this publication are the property of The American Ceramic Society. Any duplication, reproduction, or republication of this publication or any part thereof, without the express written consent of The American Ceramic Society or fee paid to the Copyright Clearance Center, is prohibited.

properties at low temperatures. Such techniques are compatible with many semiconductor fabrication technologies.[8]

Thin films of SBT have been synthesized using a mixed acetate-alkoxide route designed by the authors. Starting materials for this route include Bi acetate and a Sr-Ta double-ethoxide. In this paper, the effects of heat treatment and composition on the microstructure and ferroelectric properties of SBT films deposited using this approach are presented.

PROCEDURE

Fig. 1 is a flow chart outlining a typical film fabrication process.

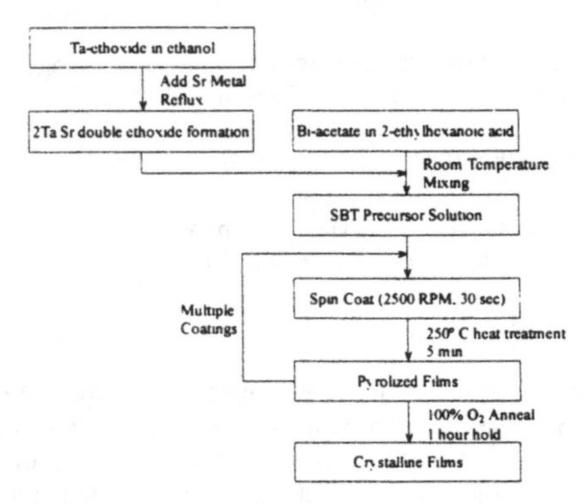

Figure 1: Schematic of SBT film fabrication.

Ta-ethoxide (99.99% Strem) and Sr-metal (99% Alfa) are mixed together in dry ethanol and refluxed (~80 C), in a dry argon atmosphere overnight to form a Ta-Sr double-ethoxide. The mixed alkoxide is used because Sr-ethoxide is inherently unstable in ethanol and will, with time, precipitate out of solution. Use of the soluble double-ethoxide thus facilitates the incorporation of Sr into the ethanol solution.

Bi-acetate was used as the Bi source for the films. Bi-acetate is preferred over the ethylhexanoate and nitrate due to its higher purity and lower adsorbed water. Bi-alkoxides are not viable sources of Bi, because they are insoluble in simple alcohols. Alkoxyalcohols may be used to dissolve the Bi-alkoxides, but the associated health hazards of alkoxyalcohols limit their use. In addition, Bi-alkoxides produced by the reaction of $BiCl_3$ and Na-alkoxide[9] often contain small quantities of the reaction by-product, NaCl. Salt contamination can compromise the electrical properties of films because the high mobility of the salt increases the dielectric loss ($\tan\delta$). To purify Bi-alkoxides, these contaminants can be removed using expensive vacuum distillation processes. In comparison, high purity Bi-acetate is inexpensive and readily available.

The coating solution was made by dissolving, in separate flasks, the Ta-Sr double-ethoxide in ethanol and Bi-acetate (99.99% Aldrich) in 2-ethylhexanoic acid (Alfa). The Bi-acetate solution is then added to the double-ethoxide solution and allowed to mix at room temperature for 15-20 min. The molarity of the resulting solution was 0.15 mol SBT/liter.

The precursor solution was then removed from the flask using a syringe and filtered through 0.2 µm filters onto a silicon wafer containing $Pt/TiO_x/SiO_2/Si$ (100) electrode stacks. The flooded wafer was spun at 2500 RPM for 30 seconds to form a thin coating. After each coating, the wafer was heated to 250 C on a hot plate, for approximately five min., to evaporate any remaining ethanol and

2-ethylhexanoic acid, and to partially decompose the organic components. The coating process was repeated (5-6 times) to reach the targeted green thickness of 0.3-0.4 μm. A final heat treatment at 750 or 830 C (2.5 C/min heating rate) for one hour in flowing O_2 was used to crystallize the films. Post-firing thicknesses ranged from 0.18-0.35 μm. Capacitors were fabricated using shadow masking to deposit Pt electrodes with cross-sectional areas of ~2E-3 cm^2 onto the films. A post-metallization anneal of one hr. at 325 C in O_2 was performed prior to electrical testing.

The phase development of the films was monitored by X-ray diffraction using Cu Kα radiation. Microstructural characterization was performed with a Hitachi S-4500 Field-emission Scanning Electron Microscope operating at an accelerating voltage of 15 kV. Electrical characteristics at 120 Hertz were determined using the Radiant Technologies RT66A Ferroelectric Tester in virtual ground mode. Film thickness was measured with a Dektak II profilometer.

RESULTS AND DISCUSSION

Phase Development

The XRD patterns for SBT films with various compositions fired for one hr. at 750 C are shown in Fig. 2. The number sequence shown next to the XRD patterns on the graph and referred to in the text are the target compositions. The first number is the number of moles of Sr available to make one mole of SBT; any number less than 1 represents a Sr deficient solution. The second number is the number of moles of Bi available to make one mole of SBT, so any number greater than 2 depicts a Bi rich solution. The third number is the number of moles of Ta available to make one mole of SBT. The compositions shown in Fig. 2 all have 10% excess Bi and, with respect to the Sr content, are stoichiometric (1/2.2/2), 10% deficient (0.9/2.2/2), 20% deficient (0.8/2.2/2), and 30% deficient (0.7/2.2/2) in Sr.

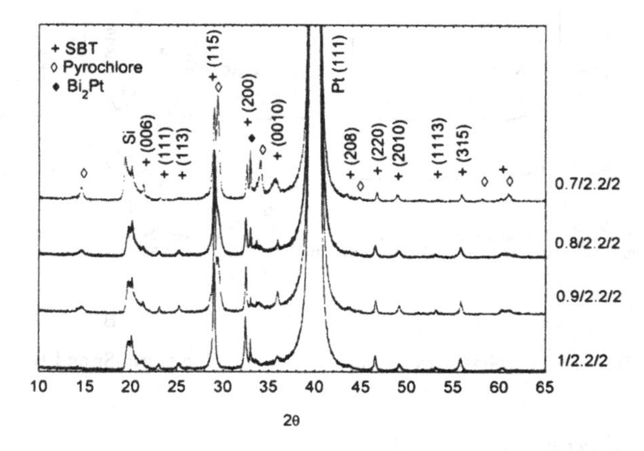

Figure 2: Effect of Sr deficiency on phase development for SBT films fired at 750 C.

The XRD patterns indicate that the layered perovskite phase is the major phase in all of the samples. Many of the films contain pyrochlore ($Sr_{0.2}(Sr_{0.5}Bi_{0.7})Ta_2O_{6.75}$) as a second phase.[3] The amount of pyrochlore increases as the Sr content decreases. The pyrochlore phase is identified by the presence of diffraction peaks at 14.5°, 29.4°, 35.9° and 61° 2-theta, based on theoretical patterns calculated by Rodriguez *et. al* [3] and a comparison with internally produced standards.

The compositional dependence of the intensities of the pyrochlore peaks is consistent with XRD patterns published by Hase *et. al*[10] and Atsuki *et. al*[11]. Noguchi *et. al*[12] also found a similar trend, but identified their second phase as $BiTaO_4$. The locations of the second phase peaks shown in Fig. 2 are not consistent with those reported on the JCPDS pattern for $BiTaO_4$ (16-909).

It seems odd that pyrochlore formation is so prevalent in Sr-deficient films, especially since the films included excess Bi. Pyrochlore is typically formed in Bi-deficient films, as discussed by Rodriguez *et. al.*[3, 13] A possible explanation for the increase in pyrochlore formation is that the mobility of Bi atoms is enhanced due to the increase in Sr vacancies with increased Sr deficiency. The XRD patterns clearly show that the volume fraction of Bi_2Pt increases with Sr deficiency, which suggests that a larger amount of Bi may be diffusing to the bottom Pt electrode. It is possible that the increased formation of Bi_2Pt may eventually deplete the film of Bi, thereby promoting the formation of pyrochlore. This is supported by the fact that the stoichiometric Sr composition (1/2.2/2) shows no evidence of pyrochlore formation and less Bi_2Pt than the highly Sr-deficient films. XRD patterns for 830 C show a similar trend to that observed for the 750 C films, save that the intensities of the diffraction peaks associated with the second phases are higher.

Figs. 3A-B show the effect of Bi content on the phase development for stoichiometric Sr and 20% Sr deficient films at 750 C. The XRD patterns show that variations in Bi content over the range of this study (stoichiometric to 20% excess) do not substantially effect the phase development (for a given Sr content). Phase-pure SBT was formed in stoichiometric Sr films and a mixture of pyrochlore and SBT was present in 20% Sr deficient films.

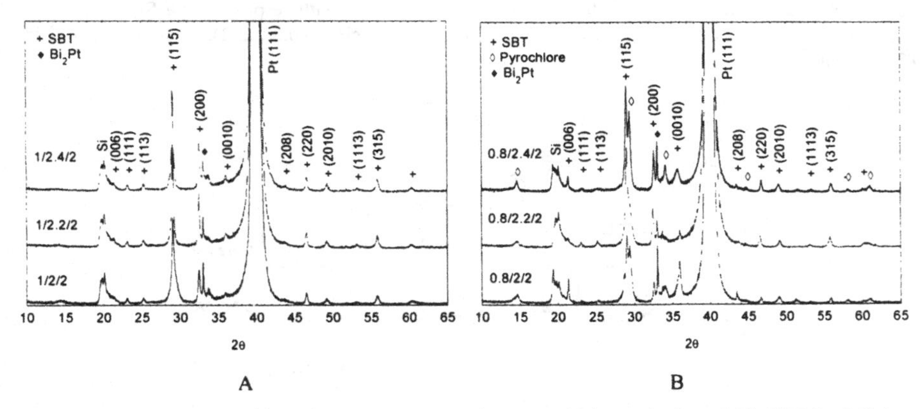

Figure 3: Effect of Bi excess on phase development for A) stoichiometric Sr and B) 20% Sr deficient films fired at 750 C.

Electrical Properties/Microstructure

Figs. 4A-B summarize the effects of Sr deficiency on the electrical properties of SBT films fired at 750 C and 830 C under an applied field of 250 kV/cm.

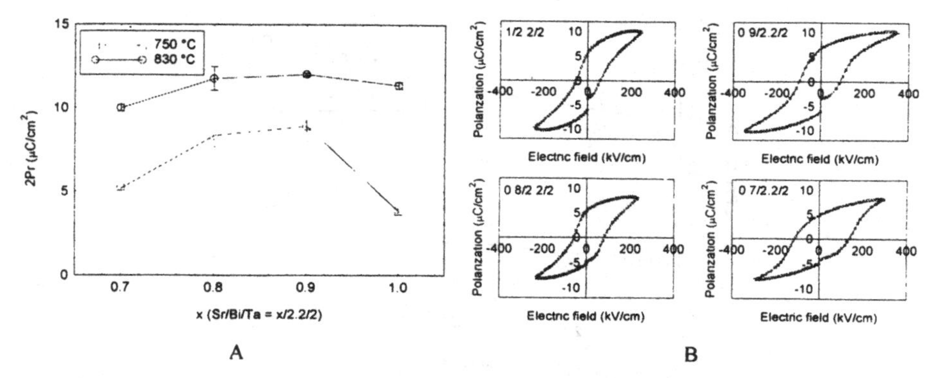

Figure 4: Effect of Sr composition on the A) 2Pr and B) Hysteresis behavior of SBT films.

Despite the presence of secondary phases, Sr-deficient films have much higher values of 2Pr (the sum of the absolute values of the magnitudes of the remanent polarizations after the removal of equal fields of opposite signs) than stoichiometric films fired at 750 C (8.7 $\mu C/cm^2$ at 10% Sr deficient vs. 3.8 $\mu C/cm^2$ for stoichiometric films). The lower value of 2Pr observed in the stoichiometric films can likely be attributed to the finer scale microstructure. This is supported by the fact that after firing at 830 C, the 2Pr values of the stoichiometric Sr composition films were only slightly lower than those of the Sr-deficient films (12±0.2 $\mu C/cm^2$ at 10% Sr deficient vs. 11.3±0.4 $\mu C/cm^2$ for stoichiometric films).

Microstructural analysis showed that the grain size and overall morphology of the stoichiometric Sr films were almost identical to those of the Sr-deficient films after the 830 C heat treatment. In both the 750 C and 830 C film series, the maximum observed 2Pr occurred between 10% and 20% deficient Sr, which is consistent with the literature.[10-12]

The increased 2Pr is not without a liability. As Fig. 4B shows, the coercive field (2Ec) of the films (under an applied field of 250 kV/cm) increases with Sr deficiency from 100 kV/cm for stoichiometric Sr films to 250 kV/cm for 30% deficient films. The increase in the coercive field is likely caused by the presence of the pyrochlore phase, which may pin domain walls under low electric fields. This has significant implications for commercial applications in which low driving voltages limit the magnitude of the coercive field.

The Bi content was also found to have a significant effect on the electrical properties of the films, as shown in Fig. 5. Unlike the Sr content, which controlled the amount of second phase formation, Bi content appears to affect primarily the development of the microstructure. Regardless of the Sr content, the values of 2Pr shown in Fig. 5 reach a maximum at 10 % excess Bi. However, the electrical properties vary widely between the different compositions and heat treatments.

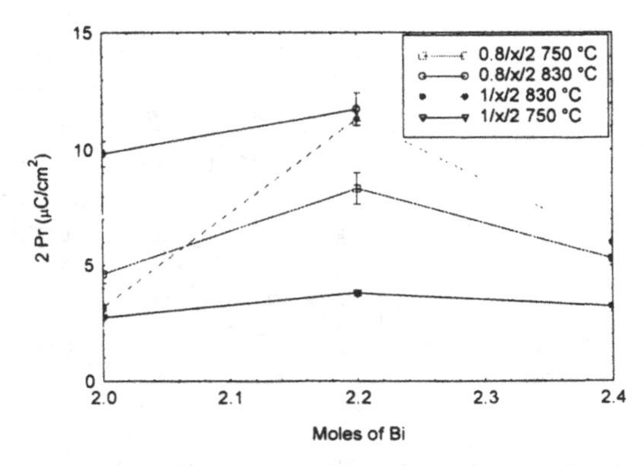

Figure 5: Effect of Bi Excess on the coercive field (2Ec) of SBT films.

Analysis of the microstructures of the stoichiometric Sr films suggests that differences in grain size probably account for the observed differences in the measured values of 2Pr. For stoichiometric Sr films with 10% excess Bi, the average grain size is approximately 70 nm, compared to 60 nm in stoichiometric Bi and 20% excess Bi films fired at 830 C (see Fig. 6). The grain sizes for the stoichiometric Sr films are all approximately 50 nm at 750 C, which explains the similarity in electrical properties at that temperature.

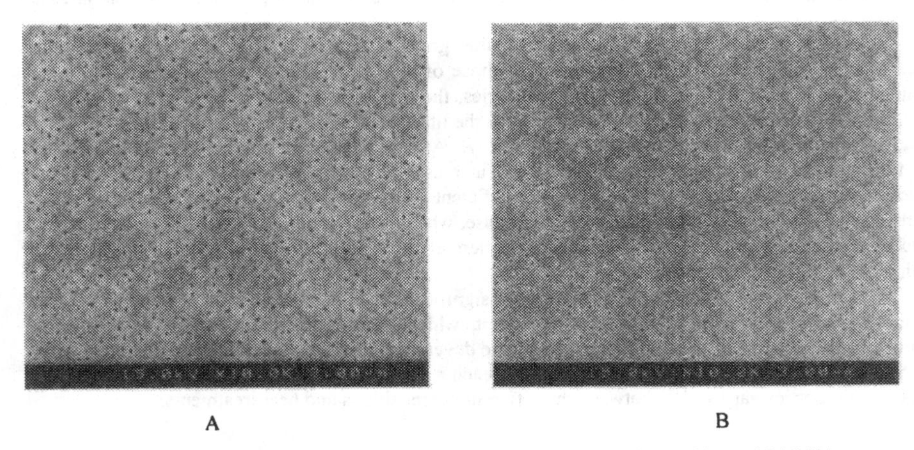

Figure 6: Representative microstructures of stoichiometric Sr films with A) 10% and B) 20% excess Bi fired at 830 C.

Sr-deficient films have a much different microstructure than that of the stoichiometric Sr films. The 20% Sr deficient, 10% excess Bi films have a microstructure similar to stoichiometric Sr films, with average grain sizes of 70 nm and 85 nm at 750 C and 830 C, respectively. However, when the Bi content is either stoichiometric or 20% excess, the microstructure dramatically changes. Fig. 7 shows representative microstructures of 20% Sr deficient films with 10% and 20% excess Bi fired at

830 C. Instead of a continuous film of small grains, the microstructure at 20% excess Bi consists of very large grains (1-3 μm) of SBT separated by extremely fine grains of pyrochlore (based on the XRD results) ~40 nm in size.

It is believed that the presence of such a large volume fraction of non-ferroelectric second phase counters the benefits of the enhanced SBT grain size by reducing the total amount of ferroelectric layered-perovskite SBT, within a given capacitor. Alternately, the grain size may only have a significant positive influence on 2Pr up to a specific grain size, and an increase over that value has a minimal effect on improving the ferroelectric properties. The results of Fig. 8 illustrate the effect of grain size in controlling ferroelectric properties. In this figure, values of 2Pr are plotted as a function of grain size for several Sr-deficient and Bi-rich films. The results suggest that composition and firing temperature are important only in so far as they determine the grain size of the final microstructure. Current research efforts are focused on testing the validity of this assertion.

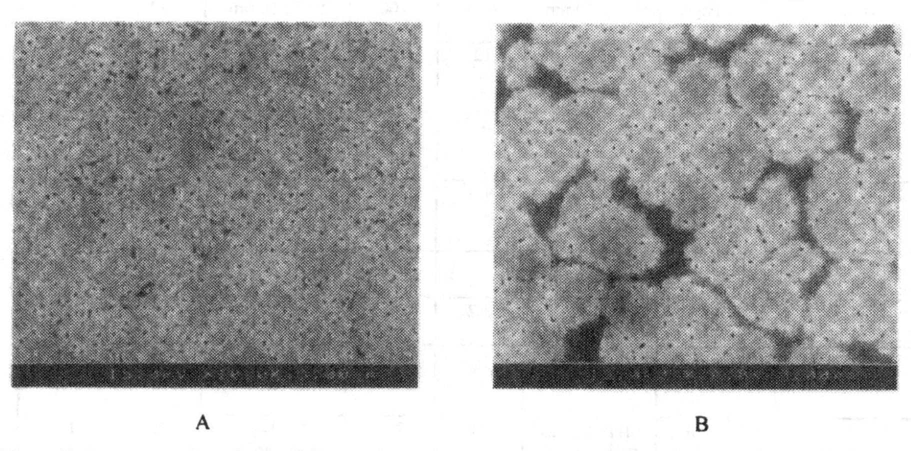

Figure 7: Representative microstructures of 20% Sr deficient films with A) 10% and B) 20% excess Bi fired at 830 C.

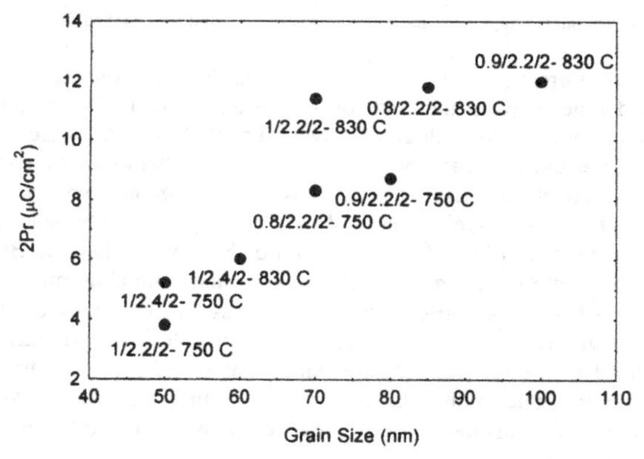

Figure 8: Effect of grain size on the 2Pr of SBT films.

SUMMARY AND CONCLUSIONS

In this work, a new alkoxide-acetate route for fabricating Sr-Bi tantalate (SBT) thin films has been described. The new route offers a way of fabricating SBT films with non-hazardous and relatively inexpensive precursor materials. Table I summarizes the results of compositional studies designed to examine the effects of Sr deficiency and Bi excess on phase development, microstructure, and electric properties of thin films produced via this new route.

Table I: Composition Study Results

Composition (Sr/Bi/Ta)	Phases	Grain Size (830 C)	2Pr (830 C)	2Ec (830 C)	Grain Size (750 C)	2Pr (750 C)	2Ec (750 C)
1/2.2/2	SBT. Bi_2Pt	SBT: 70 nm	11.4	100	SBT: 50 nm	3.8	60
0.9/2.2/2	SBT, Bi_2Pt, Pyrochlore	SBT: 100 nm	12	185	SBT: 80 nm	8.7	105
0.8/2.2/2	SBT. Bi_2Pt, Pyrochlore	SBT: 85 nm	11.8	150	SBT: 70 nm	8.3	120
0.7/2.2/2	SBT, Bi_2Pt. Pyrochlore	SBT: 1-3 μm Pyro: 40 nm	9.9	250	SBT: 1-3 μm Pyro: 40 nm	4.3	150
1/2.4/2	SBT. Bi_2Pt	SBT: 60 nm	6	100	SBT: 50 nm	5.2	85
1/2/2	SBT. Bi_2Pt	-	3.2	130	-	2.7	95
0.8/2.4/2	SBT. Bi_2Pt, Pyrochlore	SBT: 1-3 μm Pyro: 40 nm	-	-	SBT: 1-3 μm Pyro: 40 nm	5.3	140
0.8/2/2	SBT. Bi_2Pt. Pyrochlore	SBT: 1-3 μm Pyro: 40 nm	9.8	225	SBT: 1-3 μm Pyro: 40 nm	5.4	150

Note: The 2Pr and 2Ec values are quoted in $\mu C/cm^2$ and kV/cm, respectively (under an applied field of ~250 kV/cm).

The results show that the formation of second phases (identified as pyrochlore and Bi_2Pt) increases with an increase in Sr deficiency (over the range of 0-30 % deficiency). Stoichiometric Sr films were mostly phase-pure layered perovskite, with a small fraction of Bi_2Pt second phase. Bi excess, over the range of 0-20% excess, was found to have no significant effect on the phase development of the SBT films, but it did have an effect on microstructural development. Stoichiometric Sr films with 10% excess Bi were found to have larger grain sizes, and subsequently higher 2Pr values, than films with 0 and 20% excess Bi. Sr-deficient (20%) films showed a dichotomy. When the Bi content was 10% excess, the resulting films were quite similar to the stoichiometric films in microstructure (unimodal grain size distribution); but, stoichiometric Bi and 20% excess Bi films showed a bimodal grain size distribution, with very large SBT grains surrounded by fine pyrochlore second phase.

When correlated with electrical measurements, analysis of the microstructure revealed that the grain size may be the determining factor for establishing the magnitude of the remanent polarization of the films. Composition and heat treatment may be important only in so far as they establish a particular grain size.

In terms of hysteresis behavior, the 10% excess films had higher 2Pr values than the stoichiometric Bi and 20% excess Bi films. It is believed that the presence of the large volume fraction of non-ferroelectric second phase counters the benefits of the enhanced SBT grain size by reducing the total amount of ferroelectric layered-perovskite SBT within a given capacitor. It may also be possible that the grain size only has a significant positive influence up to a specific grain size, and that an increase over that value has minimal impact on the ferroelectric properties. Hysteresis measurements show that the ideal composition for the maximum 2Pr is between 10% and 20% Sr deficiency with 10% excess Bi, where the 2Pr is as high as 12 $\mu C/cm^2$ under an applied field of 250 kV/cm. However, it was found that the coercive field also increases with increasing Sr deficiency. Therefore, a compromise may need to be made in order to fulfill the 2Pr and 2Ec requirements of commercial applications.

Brian J. J. Zelinski and Donald R. Uhlmann are the recipients of a Motorola Semiconductor Products Sector Sponsored Research Project.

REFERENCES

1. B. Aurivillius, "Mixed bismuth oxides with layered lattices I. The structure of $CaNb_2Bi_2O_9$," *Arkiv For Kemi* **1**, 463-480 (1949).
2. V. A. Isupov, "Properties of solid solutions of $A_{(m-1)}Bi_2M_{(m)}O_{(3m+3)}$ layered perovskites," *Phys. Solid State* **37**, 166-169 (1995).
3. M. Rodriguez, T. J. Boyle, B. A. Hernandez, C. D. Buchheit, and M. O. Eatough, "Formation of $SrBi_2Ta_2O_9$: Part II. Evidence of a bismuth-deficient pyrochlore phase," *J. Mater. Res.* **11**, 2282-2287 (1996).
4. P. Y. Chu, R. E. Jones. Jr., P. Zurcher, D. J. Taylor, B. Jiang, and S. J. Gillespie, "Characteristics of spin-on ferroelectric $SrBi_2Ta_2O_9$ thin film capacitors for ferroelectric random access memory applications," *J. Mater. Res.* **11**, 1065-1068 (1996).
5. B. Jaffe, W. R. Cook Jr., and H. Jaffe, *Piezoelectric Ceramics*. J. P. Roberts and P. Popper, Eds., Non-Metallic Solids (Academic Press, London, 1971), vol. 3.
6. D. F. Nelson, Ed., *Low Frequency Properties of Dielectric Crystals*, vol. 29 (Springer-Verlag, Berlin, 1993).
7. A. D. Rae, "Structure refinement of commensurately modulated bismuth strontium tantalate, $Bi_2SrTa_2O_9$," *Acta. Cryst.* **B48**, 418-428 (1992).
8. B. A. Tuttle and R. W. Schwartz, "Solution deposition of ferroelectric thin films," *MRS Bulletin* **21**, 49-54 (1996).
9. C. D. Gutleben, Y. Ikeda, C. Isobe, A. Machida, T. Ami, K. Hironaka and E. Morita, "The microstructure of $SrBi_2Ta_2O_9$ films," S. B. Desu, D. B. Beach, P. C. V. Buskirk, Eds., Metal-Organic Chemical Vapor Deposition of Electronic Ceramics II, Boston, MA (MRS, 1996).
10. T. Hase, T. Noguchi, K. Amanuma and Y. Miyasaka, "Sr content dependence of ferroelectric properties in $SrBi_2Ta_2O_9$ thin films," *Integrated Ferroelectrics* **15**, 127-135 (1997).
11. T. Atsuki, N. Soyama, T. Yonezawa and K. Ogi, "Preparation of Bi-based ferroelectric films by sol-gel method," *Jpn. J. Appl. Phys.* **34**, 5096-5099 (1995).
12. T. Noguchi, T. Hase and Y. Miyasaka, "Analysis of the dependence of ferroelectric properties of strontium bismuth tantalate (SBT) thin films on the composition and process temperature," *Jpn. J. Appl. Phys. Part 1* **35**, 4900-4904 (1996).
13. M. A. Rodriguez, T. J. Boyle, C. D. Buchheit, R. G. Tissot, C. A. Drewien, B. A. Hernandez and M. O. Eatough, "Phase formation and characterization of the $SrBi_2Ta_2O_9$ layered-perovskite ferroelectric," *Integrated Ferroelectrics* **14**, 201-210 (1997).

THE ROLE OF PROCESS VARIABLES ON MICROSTRUCTURAL DEVELOPMENT IN SOL-GEL DERIVED SBN THIN FILMS

A.Y. Oral and M. L. Mecartney
University of California, Irvine Department of Chemical and Biochemical Engineering and Material Science Irvine, CA 92697-2575

ABSTRACT

Microstructural changes in sol-gel derived $Sr_xBa_{1-x}Nb_2O_6$ (SBN) thin films were monitored as a function of Ba to Sr ratio, substrate (Si, AlN or MgO), and processing variations. Sols were created using Ba, Sr and Nb alkoxides dissolved in alcohol and acetic acid. The tetragonal tungsten bronze structure was achieved in Sr rich films (SBN73) at 750°C. Films with equal Ba:Sr amounts (SBN50) resulted predominantly in the formation of the orthorhombic crystal structure. High decomposition temperatures of the organics led to a tendency to form a porous microstructure, but careful control of thermal process parameters could be used to optimize film microstructures.

INTRODUCTION

Strontium barium niobate ($Sr_xBa_{1-x}Nb_2O_6$, SBN) is a ferroelectric solid solution formed between $BaNb_2O_6$ and $SrNb_2O_6$. The region of solid solution for the tungsten bronze structure is $0.25 < x < 0.75$[1]. Two different lattice types and symmetries have been observed in solid solution. The first one is orthorhombic, with space group Cmm2, having approximate lattice parameters a,b = 1.7 nm, c = 0.78 nm. The second one is tetragonal, with space group P4bm and approximate lattice parameters of a,b =1.2 nm, c=0.39 nm[2]. A change from the tetragonal structure to the orthorhombic structure has been observed for the values of $x < 0.60$ for SBN powders[3].

The basic formula for a ferroelectric tetragonal tungsten bronze is $[(A1)_2 (A2)_4 C_4] [(B1)_2 (B2)_8] O_{30}$[4]. The skeletal framework of the tungsten bronze structure is formed by MO_6 octahedra, which share corners to form the cavities of A1, A2 and C[2]. When two cations are found in the tetragonal bronze structure, the smaller one prefers A1 site because of relatively small size of A1 sites compared to A2 sites. The C is the smallest of the three cavities created by the framework of octahedra. Cations which commonly occupy the C site are lithium,

To the extent authorized under the laws of the United States of America, all copyright interests in this publication are the property of The American Ceramic Society. Any duplication, reproduction, or republication of this publication or any part thereof, without the express written consent of The American Ceramic Society or fee paid to the Copyright Clearance Center, is prohibited.

magnesium and beryllium. Jamieson[5] proposed that when different types of A and C cations are added to the structure, tilting of MO_6 octahedra causes the symmetry to change from tetragonal to orthorhombic. The atoms involved in these shifts are O atoms which are located at the apex of the corner and edge octahedra of the orthorhombic cell for SBN.

Jamisons[5] atom distribution model is still accepted for SBN where Ba cations are only found in large A2 sites and Sr atoms are preferably occupied in the small A1 sites. Nevertheless, a small percentage of Sr atoms also occupy A2 sites especially in the Sr rich solid solutions.

SBN is a ferroelectric material whose Curie temperature increases from 60 to 250°C as x decreases from 0.75 to 0.25. At the Curie temperature all the cations are displaced offset from the planes of oxygen atoms along the c axis resulting in a P_s value of 6.4 μC/cm in the [001] direction. As the Ba content of solution increases, SBN shifts from relaxor ferroelectric to a normal ferroelectric. During this transition, as the Sr content decreases, A1 sites become less occupied and Sr/Ba atomic ratio in the A2 sites decreases. In addition the a, and b lattice parameters increase, possibly indicating that NbO_6 octahedra are shifting from a tilted to a more ordered arrangement. As Ba increases and reaches 40%, this shift possibly causes the structure to become orthorhombic. These local structural distortions may be responsible for broadening of the temperature range for the paraelectric to ferroelectric transition leading to relaxor ferroelectric behavior.

$Sr_xBa_{1-x}Nb_2O_6$ thin films are candidates as potential materials for pyroelectric infrared detectors, piezoelectrics, electro-optics, and photorefractive optics[6]. There are four major deposition techniques that have been successfully applied to prepare films. Among them sol-gel processing is less complex and inexpensive compared to sputtering, MOD, and CVD. In addition, sol-gel processing has the advantages of excellent homogeneity, ease of chemical composition control, and low processing temperature which makes it a very convenient method for thin film preparation.

This present study uses a new solvent to make the sols less susceptible to hydrolysis and precipitation than alcohol based solvents and investigates the role of process variables on microstructural development of the films.

EXPERIMENTAL PROCEDURES

A flow diagram of the experimental procedure is given in Figure 1. A new type of sol was prepared by dissolving Ba-isopropoxide and Sr-isopropoxide in acetic acid and mixing with Nb-ethoxide. This is unlike most previous studies of SBN sol-gel formation which used alkoxides in alcohol[6]. The final solution, which had a concentration of 0.36M, was clear light brown in color without any suspension of particles. The solutions were spun at 2000 rpm for 60s onto polycrystalline AlN, (100) Si, (111) Si, and (100) MgO substrates. 1-10 layers were deposited followed by air drying for 1 minute. Two different methods of heat treatments were used. The first was fast thermal processing (FTP). After air drying a pyrolysis heat treatment at 350°C was applied to each layer of the FTP films. Then the FTP films were inserted directly into a furnace at 750°C. The second heat treatment method was conventional furnace annealing with information from thermogravimetric analysis (TGA) experiment used to determine heating rate. Heating rates as slow as 1°C/min were applied for the regions of the TGA curves with sharp weight losses. In addition, the temperature was held for an hour at the mid- points of every major weight loss temperature. (Details will be given in the "Effect of Thermal Processing on Microstructure" section)

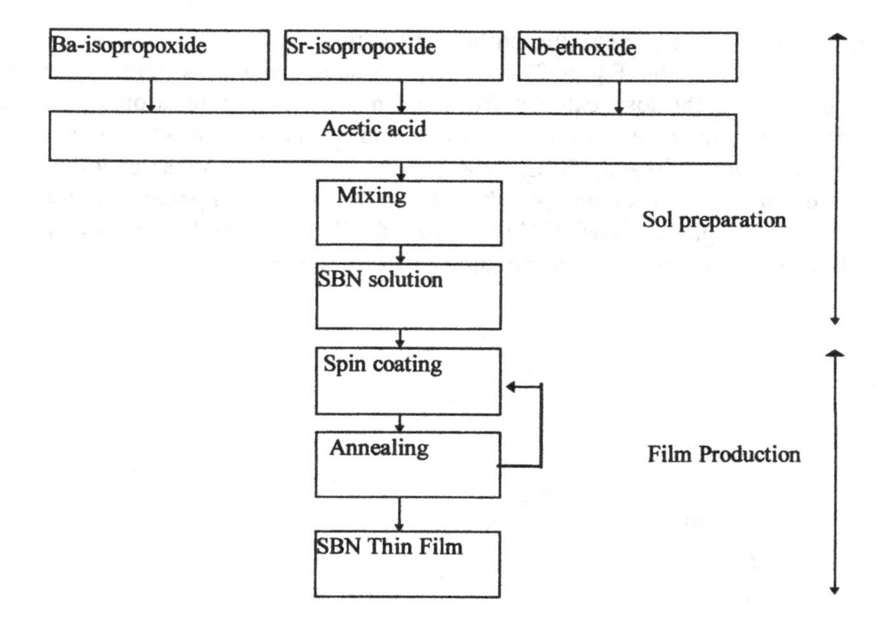

Figure 1. Flow diagram for the sol-gel processing of SBN films.

The crystal structures and crystallization behaviors of the films were analyzed by X-ray diffraction (XRD), Differential scanning calorimeter (DSC) and thermogravimetric analysis (TGA). Surface morphology of the films were examined by Scanning Electron Microscopy (SEM).

RESULTS AND DISCUSSION

Use of New Solvent

In previous studies, ethanol was used as a common solvent for the preparation of the sol-gel SBN thin films. These sols were difficult to prepare due to high sensitivity to the moisture in air, low shelf life and long preparation time of ethoxides from Sr and Ba metals[6]. This new sol preparation method using acetic acid as a main solvent eliminates these problems. Since the acetic acid is a suitable chelating agent and makes the sols less susceptible to fast hydrolysis. The new sols have very low sensitivity to moisture and a shelf life of several months. Furthermore, preparation of these new sols take only 20 minutes while ethanol based sols take 6-12 hours.

Effect of Thermal Processing on Microstructure

TGA graphs (Figure 2) indicate three steep weight loss temperature ranges. The first weight loss extends from room temperature to around 150°C, and corresponds to the evaporation of water and acetic acid. A second sharp weight loss is between 300 and 400°C. DSC graphs (Figure 3) show an endothermic peak corresponding to this range possibly caused by the low temperature evaporation or burnout of organics. The DSC peak around 550°C may be due to crystallization of SBN. Since there is no corresponding weight loss (Figure 2).

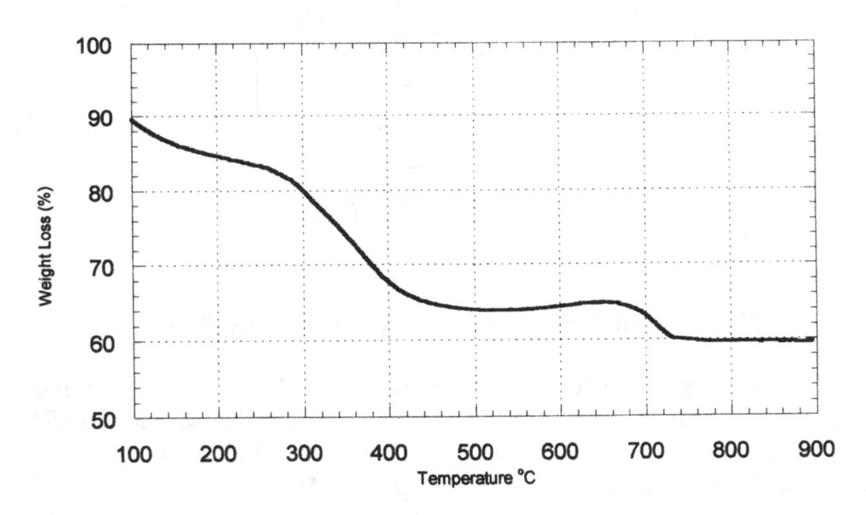

Figure 2. TGA of SBN50 Powder. Heating rate is 10°C/min.

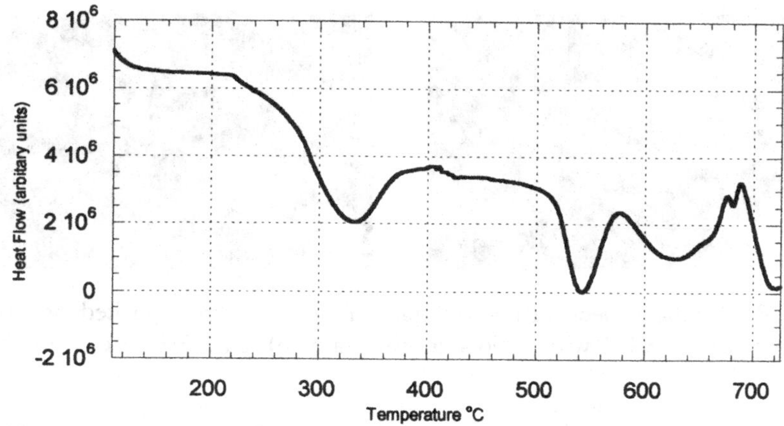

Figure 3. DTA of SBN50 Powder. Heating rate is 10°C/min.

The weight loss between the 650 and 720°C is caused by decomposition of bound organics and the combustion of free carbon in the films and confirmed by exothermic peaks in the DSC graphs around 680°C. Although there seems to be significant weight loss at high temperatures, it was found that the reactions associated with these weight losses shift to lower temperatures at slower heating rates.

SEM micrographs of SBN50 films on Si substrates exhibited substantial differences with variations of heating rates during crystallization of the films. The films that were produced by conventional furnace annealing (CFA) at a very slow heating rates were uniform and crack and pore free throughout the film (Figure 4a). On the other hand, the films produced by fast thermal annealing (FTP) were full of cracks and pores (Figure 4b). In addition, these films contained bubbles caused by entrapped gases between the substrate and the film. During FTP, the decomposition of organics at high temperatures is responsible for the pores and bubbles. Fast evaporation at low temperatures forms macroscopic holes as well. Cracks formed on the films are due to rapid shrinking of the films. Steep temperature changes during FTP enhance the crack formation. Films produced by CFA did not form macroscopic holes. Organics evaporate through pores in the film. In the later stages of the annealing, sufficient time at high temperatures allows diffusion to annihilate these small holes.

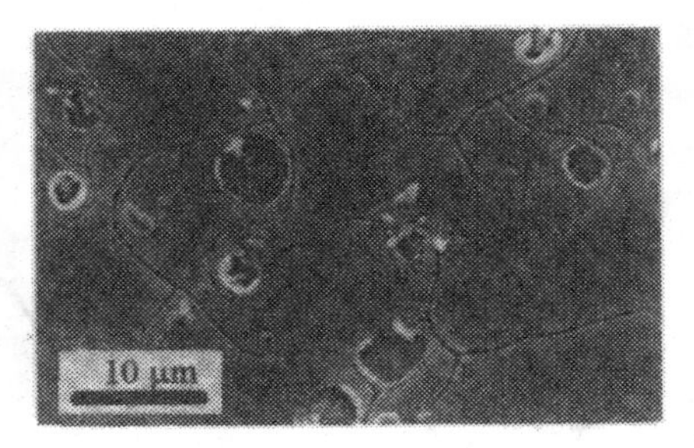

Figure 4. Scanning electron micrographs of SBN50 film deposited on (100) Si substrate (a) annealed with a slow heating rate (b) annealed with a fast heating rate.

Effect of Composition and Substrate to the Crystal Structure

Two different crystal structures were observed depending on the composition of the sols. 50/50 composition deposited on AlN had primarily orthorhombic phase (Figure 5). SBN73 sols (x=0.73) predominantly crystallized in the tetragonal tungsten bronze structure on AlN (Figure 6). This phenomenon has been seen in the bulk SBN around 1400°C[6]. Yet, the same transformation is observed at significantly lower temperatures by using sol-gel processing. 100% tungsten bronze phase was obtained in films deposited on MgO substrates (Figure 7).

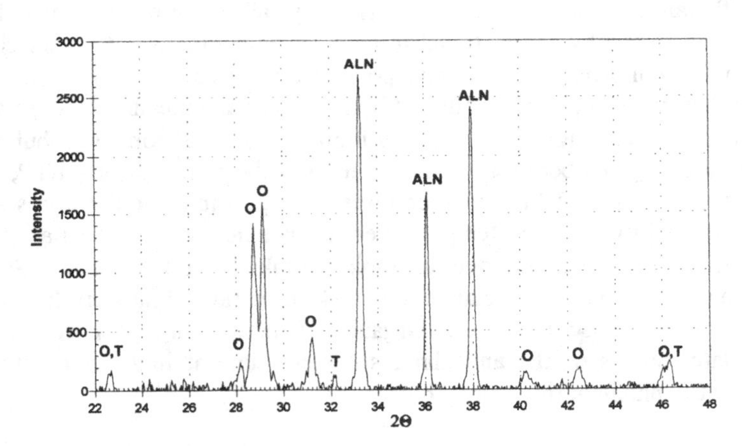

Figure 5. XRD pattern of SBN50 film with predominantly orthorhombic structure deposited on polycrystalline AlN substrates and annealed at 750 °C.

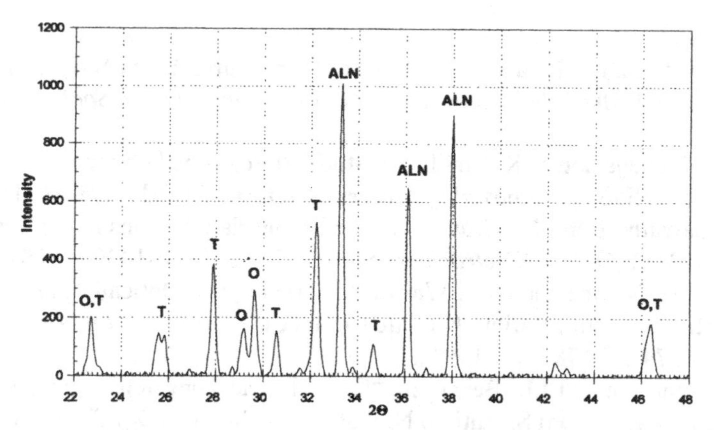

Figure 6. XRD pattern of SBN73 film with predominantly tetragonal structure deposited on polycrystalline AlN substrates and annealed at 750°C.

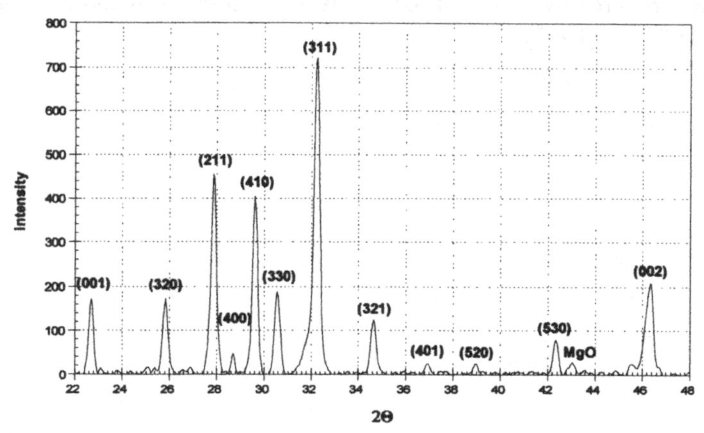

Figure 7. XRD pattern of SBN73 film with 100% tetragonal structure deposited on polycrystalline (100) MgO substrates and annealed at 750°C.

CONCLUSIONS

1. Good quality SBN thin films can be produced by sols made by dissolution of alkoxides in acetic acid.
2. The heating rate is a key aspect to the film quality. Slower heating rates allow decomposition reactions to occur at lower temperatures and produce defect free films.
3. SBN50 sols produces primarily orthorhombic films while tetragonal tungsten bronze films were obtained from SBN73 sols heat treated at 750°C.

REFERENCES

[1] W. Sakato and S. Hirano, "Synthesis of Strontium Barium Niobate Thin Films through Metal Alkoxide," *Journal of The American Ceramic Society*, 79 [9] 2283-88 (1996).

[2] M.P. Trubelja and D.K. Smith, "A Study of Positional Order in Strontium Barium Niobate," *Journal of Materials Science,* 31 [6] 1435-43 (1996).

[3] J.R. Carruthers and M. Grasso, "Phase Equilibria Relations in the Ternary System BaO-SrO-Nb_2O_5," *J. Electrochem. Soc.*, 117 [11] 1426-1430 (1970).

[4] N. S. VanDamme and S. R. Winzer, "Fabrication of Optically Transparent and Electrooptic Strontium Barium Niobate Ceramics," *Journal of the American Ceramic Society,* 74 [8] 1785-92 (1991).

[5] P.B. Jamieson and J.L. Bernstein, "Ferroelectric Tungsten Bronze-Type Crystal Structures. I. Barium Strontium Niobate $Ba_{0.27}Sr_{0.75}Nb_2O_{5.78}$," *The Journal of Chemical Physics,* 48 [11] 5048-5057.

[6] C.H. Luk and K.H. Wong, "Characterization of Strontium Barium Niobate Films Prepared by Sol-Gel Process Using 2-Methoxyethanol," *Thin Solid Films,* 298 [1] 57-61 (1997).

Porous Materials through Sol-Gel Processing

MONOLITHIC HPLC COLUMN VIA SOL-GEL ROUTE

Kazuki Nakanishi, Hiroyoshi Minakuchi, Norio Ishizuka, Naohiro Soga
Department of Material Chemistry, Graduate School of Engineering, Kyoto University, Yoshida, Sakyo-ku, Kyoto 606-8501, JAPAN
Nobuo Tanaka
Department of Polymer Science and Engineering, Kyoto Institute of Technology, Matsugasaki, Sakyo-ku, Kyoto 606-8585, JAPAN

ABSTRACT

An HPLC column composed of a single piece of monolithic silica gel has been developed. By hydrolyzing alkoxysilane in the presence of a water-soluble polymer under acidic conditions, well-defined interconnected macroporous structures due to the phase separation are formed. Subsequent solvent exchange treatment on the macroporous wet gel allows one to tailor the mesopore structure within the micron-sized gel skeletons. After the solvent removal and heat-treatment, pure silica gels having controlled macropores and mesopores - double pore structure - are obtained. With the rod-column, which has a resin clad around a rod-shaped gel monolith, the chromatographic separation is performed with much lower pressure drop and weaker dependence of the plate height on the mobile phase velocity. Espcially for solutes with higher molecular mass, the separation can be performed about an order of magnitude faster than the case of conventional particle-packed columns.

INTRODUCTION

Silica gel particles have been extensively used as a packing material of columns for high-performance liquid chromatography, HPLC[1-5]. The separation efficiency of a so called particle-packed column depends on factors such as particle size and packing state. In order to speed up the analysis without sacrificing the separation efficiency, smaller particle size is preferred. However, the size of interstices of the particles necessarily decreases with the particle size, and the pressure drop increases inversely proportional to the particle size squared. Since increasingly higher column back pressure is required to obtain a constant mobile phase velocity, there arises an instrumental limitation for the accelerated separation by columns packed with small-sized particles. So far, particles as small as 5μm have been utilized as a compromise between the performance and the speed of analysis.

To the extent authorized under the laws of the United States of America, all copyright interests in this publication are the property of The American Ceramic Society. Any duplication, reproduction, or republication of this publication or any part thereof, without the express written consent of The American Ceramic Society or fee paid to the Copyright Clearance Center, is prohibited.

Recently, with the combination of the sol-gel reaction accompanied by the phase separation and subsequent solvent exchange treatment which enhances the Ostwald ripening of finely textured wet gel matrix, double-pore silica gel monoliths have been prepared[6,7]. These gels exhibit well-defined micrometer-range pores with controllable volume fraction and median size together with narrowly distributed mesopores. The first key process is the hydrolysis and polycondensation of alkoxysilane in the presence of water-soluble polymers or other additives which can induce the phase separation to form micrometer-range co-continuous domain structures parallel to the sol-gel transition of the system. The subsequent post-gelation treatments, on the other hand, affect the nanometer-range pore distribution with which the application field of gel materials is specified.

This paper firstly describes the designing principle of double-pore silica focusing on the factors which determine the macropore and mesopore characteristics. Based on the various macropore structures, the chromatographic performance of rod-columns applied to the reversed phase HPLC is described in comparison with conventional particle-packed columns.

EXPERIMENTAL

Tetramethoxysilane (TMOS) was hydrolyzed using acetic acid as a catalyst in the presence of water soluble polymers such as poly(ethylene oxide) that induce a phase separation in the polymerization stage. In order to obtain well-defined co-continuous domains of gel- and fluid-phases, the starting composition as well as the catalyst concentration and the reaction temperature were adjusted so as to induce the sol-gel transition and the phase separation concurrently. Wet gels thus formed were aged at the same temperature as the gelation stage and then immersed into dilute aqueous solutions of ammonia for various periods at varied temperatures. Finally the gels were evaporation-dried and heat-treated at 600°C for 2h.

The macropore structure was examined by a scanning electron microscope (SEM, S-510, Hitachi Ltd., Japan) using the flat fractured surface of the gel samples. The pore size distribution was determined by a mercury intrusion (PORESIZER 9320, Micromeritics, USA) and a nitrogen adsorption (ASAP 2000, Micromeritics, USA) methods, using heat-treated gel samples.

Silica rod columns for the reversed phase HPLC was prepared by fitting a resin clad around a rod-shaped gel piece and attaching end-fitting parts on both ends. Thus prepared resin-clad column was set in a Z-module (RCM 8X10, Waters Co., USA) and was used in an HPLC apparatus. Surface derivatization was performed by circulating the solutions of silane coupling agent through an unmodified column (on-column reaction). The chromatographic evaluation was carried out with ordinary HPLC instruments as described in the literature[8,9].

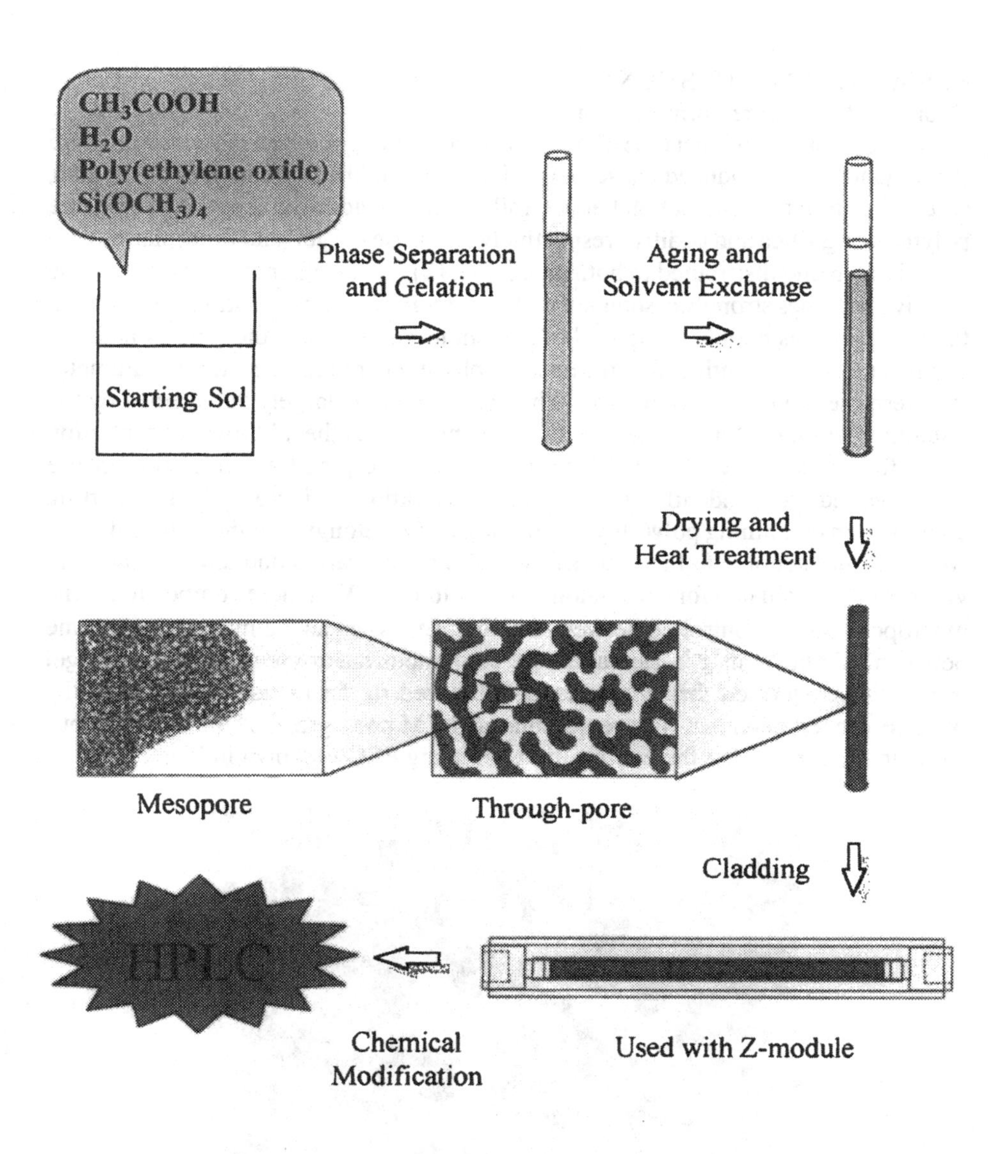

Fig. 1 Schematic flow chart of the sample preparation procedure.

RESULTS AND DISCUSSION

Choice of Macropore Forming System

There exist two distinct type of phase relations among the systems which undergo the polymerization-induced phase separation[7,10]. In the first category of systems, where the attractive interaction between silica and the additive is not so strong, the polymerizing silica and additive respectively comprise the gel- and fluid-phases with the solvent being distributed to both phases[11,12]. In the other category, where the additive becomes strongly associated with polymerizing silica to form the gel-phase, the fluid phase is mainly composed of solvent mixture. Since the ratio of additive to silica most dominantly affect the onset of phase separation, it is the key parameter to determine the median pore size. The pore volume is largely determined by the volume fraction of fluid-phase, which varies parallel to the additive concentration in the former category. In the latter category, however, the fraction of fluid-phase can be varied independently of additive to silica ratio in principle[7]. The starting composition containing poly(ethylene oxide), PEO, belonging to the latter category was therefore adopted to fabricate gel monoliths with varied domain size and pore volume to be evaluated for chromatographic columns. With these compositions, the macropore can be controlled between 0.5 to 10μm whereas the macropore volume between 0.5 and 5 cm^3g^{-1}. Hereafter, the continuous macropores formed in the gel monoliths are termed through-pores as compared to the interstitial voids in the particle-packed columns. The representative SEM photograph of the macroporous structure obtained from the composition containig PEO is shown in Fig. 2.

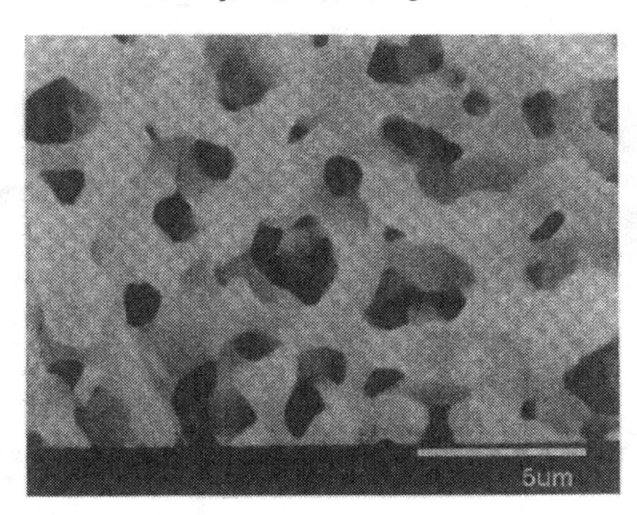

Fig. 2 SEM photograph of the typical macropore morphology of a gel sample.

Choice of Mesopore Tailoring Condition

Since the macroporous gel domains are formed already in the wet state, further tailoring of the internal pore structure by exchanging the fluid phase can be performed more efficiently and the process is less time-consuming than the case of gel monoliths without macropores. Extensive studies on the solvent exchange method revealed following points:

(1) When the pH value of the external solvent is higher than ca. 8, initially microporous gel skeletons become reorganized into those with mesopores with the median size larger than 5nm[13,14]. The in situ small-angle x-ray scattering measurements revealed that the coarsened structure is already formed in the wet state[15].

(2) The equilibrium mesopore size corresponding to the pH of the external solution can be achieved earlier as the bath ratio and/or temperature becomes higher. The distribution width tend to become broader as the temperature increases.

(3) These reorganization processes take place without essentially affecting the macroporous gel structure. Thus, the through-pores and mesopores of gel monoliths can be designed independently.

For the present series of rod columns, the wet gels after appropriate aging were immersed in a large enough amount of 0.01M aqueous ammonia solution at 40°C for 3days to produce mesopores with diameter of ca. 10nm after the heat-treatment at 600°C for 2h (Fig. 3).

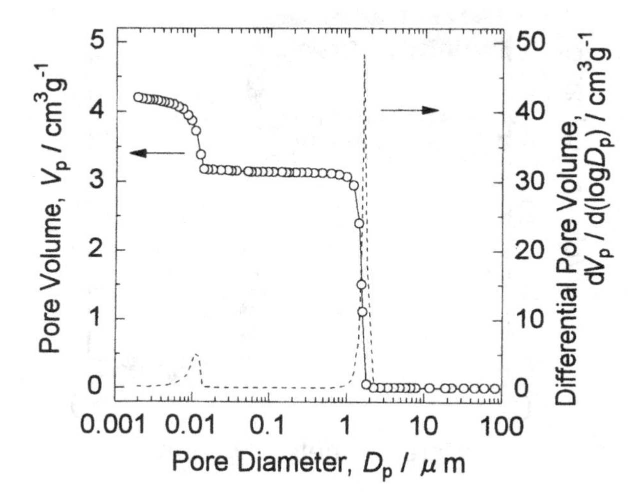

Fig. 3 Pore size distribution of typical double-pore silica prepared for chromatographic application.

Chromatographic Performance

Effect of Skeleton Thickness

In comparison with particle-packed columns in which the interparticle void space is proportional to the particle diameter, the rod columns can be designed to have thinner gel skeletons relative to the through-pore size. Table I lists the starting compositions of the gel monoliths prepared so as to retain constant through-pore sizes and varied gel-skeleton thicknesses. The macropore size distributions of respective gels are shown in Fig. 4. As the median through-pore sizes are almost constant in all the samples, those with higher pore volume have thinner skeletons.

Table I. Starting composition and macropore characteristics of heat-treated gel monoliths for rod columns with varied silica skeleton thickness.

Sample	Composition		Mesopore size	Skeleton size	Through-pore size
	PEO (g)[a]	TMOS (g)[a]	(nm)	(μm)	(μm)
SR-I	11.6	41.2	14.3	1.00	1.86
SR-II	10.2	46.4	13.1	1.16	1.73
SR-III	8.8	51.5	14.2	1.34	1.65
SR-IV	7.0	56.7	15.8	1.59	1.53

a) Mixed with 100ml of 0.01M aq soln of acetic acid at 0°C and gelled at 40°C.

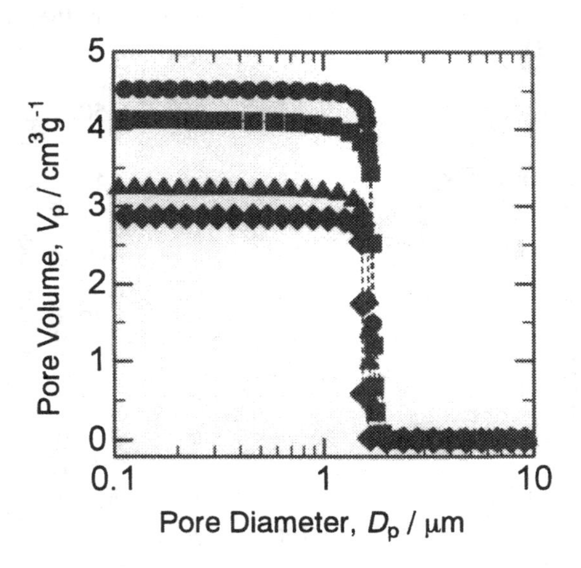

Fig. 4 Macropore size distribution of heat-treated gel monoliths obtained from the starting compositions listed in Table I. ●: SR-I, ■: SR-II, ▲: SR-III, ◆: SR-IV.

Table II. Volumes of through-pores and mesopores in a silica rod and a particle-packed column.

Volume fraction of pores in a column (%) [a]					
	Silica rod		Particle-packed column		
	SR-I		Develosil[b]		Capcellpak C_{18} UG[c]
	Before ODS	After ODS	Before ODS	After ODS	
Total porosity	86	81	78	66	60
Through-pore[d]	65	65	39	39	31
Mesopore	21	16	40	27	29
Bonded phase	--	5	—	13	--

a) Measured by size-exclusion chromatography.
b) Develosil (Nomura Chemical, Seto, Japan; particle size: 5μm, pore size: 11nm)
c) Capcellpak C_{18} UG (Shiseido, Tokyo, Japan; particle size: 5μm, pore size: 12nm)
d) Interstitial void volume in the case of particle-packed column.

Comparisons in terms of pore volume is made in Table II between the rod-column with the thinnest skeleton and a commercially available typical particle-packed columns. The rod column after ODS exhibit a through-pore volume being higher than those of particle-packed columns by factors of 1.5-2.0. Figure 5 shows the dependence of theoretical plate height for the elution of amylbenzene on the mobile phase velocity (van Deemter plot)[16]. The steep slope in *H-u* relation for the particle-packed column implies that the analytical performance becomes worse when a higher mobile phase velocity is applied. This is due to the slow diffusional motion of the solute within the particles relative to the mobile phase velocity. By contrast, all the rod columns exhibited weaker dependence and the overall slope became less steep as the skeletons became thinner.

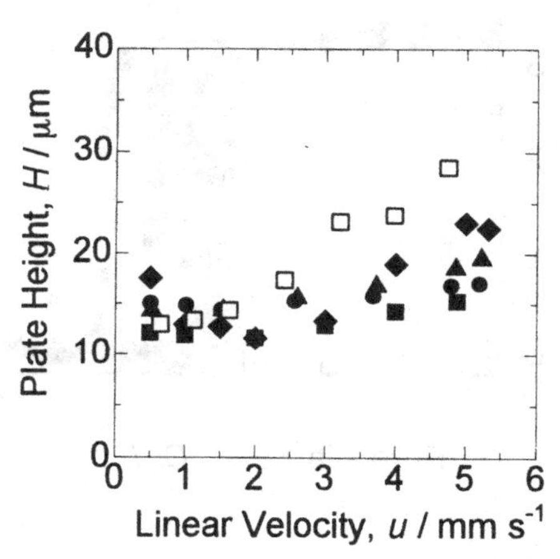

Fig. 5 Dependence of theoretical plate height, *H*, on mobile phase velocity, *u*.
●: SR-I, ■: SR-II, ▲: SR-III, ◆: SR-IV, □: Capcellpak C_{18} UG.

Effect of Domain Size

Thinner skeletons and larger through-pores in rod columns than those in particle-packed columns enable one to increase the mobile phase velocity without sacrificing the analytical performance. It is also important to see how the column efficiency varies with the domain size, defined by the sum of skeleton thickness and through-pore diameter, under constant skeleton to through-pore size ratio. Table III lists the starting compositions of the gel monoliths prepared so as to retain constant skeleton to through-pore size ratio and varied domain size between ca. 2 and 6μm. The macropore size distributions of respective gels are shown in Fig. 6.

Table III. Starting composition and resultant macropore characteristics of gel monoliths for rod columns with varied domain size.

Sample	Composition PEO[a] (g)	TMOS[a] (g)	Mesopore size (nm)	Skeleton size (μm)	Through-pore size (μm)
SR-A	9.4	46.4	15.7	2.31	3.46
SR-B	9.8	46.4	15.8	1.58	2.23
SR-C	10.2	46.4	13.1	1.16	1.73
SR-D	10.4	46.4	15.5	1.06	1.26

a) Mixed with 100ml of 0.01M aq soln of acetic acid at 0°C and gelled at 40°C.

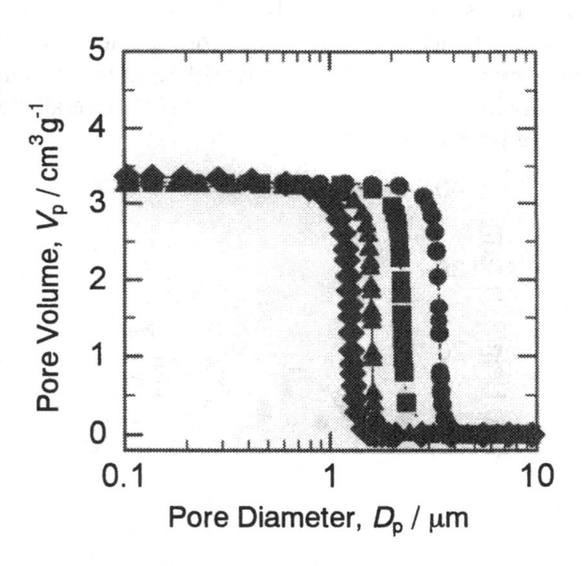

Fig. 6 Macropore size distribution of heat-treated gel monoliths obtained from the starting compositions listed in Table III. ●: SR-A, ■: SR-B, ▲: SR-C, ◆: SR-D.

As shown in Table IV, the relative volume of mesopores to skeleton and its reduction ratio after ODS are about 60% and 25% respectively for all the columns. This indicates the similarity in chemical structure of mesopore surfaces in rods and particle-packed columns. The variation in the chromatographic behavior is, therefore, mainly attributed to the difference in the macroporous morphology. The van Deemter plots for amylbenzene are shown in Fig. 7. The SR-A sample with the average skeleton thickness of 2.3μm exhibits similar results to that of 5μm particles while the SR-D sample with 1.0μm skeleton behaves similarly to 2μm particles (TSKgel Super ODS). These indicate that the rod columns having ca. 60% of through-pore volume and the skeleton thickness r give comparable chromatographic performance to that of columns packed with particles having the diameter $2r$. This difference in the sizes of skeletons and particles which give comparable performances can be partly accounted for by considering the surface to volume ratio of cylindrical and spherical geometery of gels. Between SR-A and Capcellpak, the column back pressure of the former can be 1/10 that for the latter to attain the same mobile phase velocity. Consequently, the separation impedance, E, becomes substantially lower for the former column. The separation impedance is defined as follows;

$$E = (\Delta P/N)\,(t_0/N)\,(1/\eta) \qquad (1)$$

wherer ΔP: column back pressure, N: number of plates, t_0: unit time, η: mobile phase viscosity. Similar tendencies in the van Deemter plot and in the mobile-phase velocity dependence of separation impedance have been confirmed also when higher molecular weight molecules such as insulin were used as solutes[16,17].

Table IV. Volumes of through-pores and mesopores in a silica rod and a particle-packed column.

Pore volume	Volume fraction in a column (%)[a]					
	Silica rod				Particle-packed column	
	SR-C		SR-D		Develosil[b]	
(ODS)	Before	After	Before	After	Before	After
Total porosity	86	80	86	81	76	66
Through-pore[b]	62	63	61	62	39	39
Mesopore	24	18	25	19	37	27
Skeleton Volume[c]	38	--	39	--	61	--
Silica Volume	14	--	14	--	24	--

a) Measured by size-exclusion chromatography.
b) Develosil (Nomura Chemical, Seto, Japan; particle size: 5μm, pore size: 11nm)
c) Apparent volume of silica skeleton with mesopores. Volume of silica particles in the case of particle packed column.

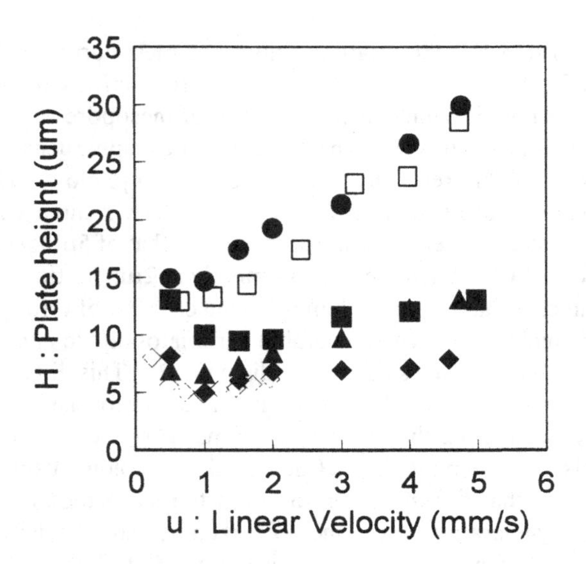

Fig. 7 Dependence of theoretical plate height, H, on mobile phase velocity, u. Solute: amylbenzene, mobile phase 80% methanol, temperature: 30°C. ●: SR-A, ▲: SR-B, ■: SR-C, ◆: SR-D, □: Capcellpak C_{18} UG (12nm), ◇: TSKgel Super ODS.

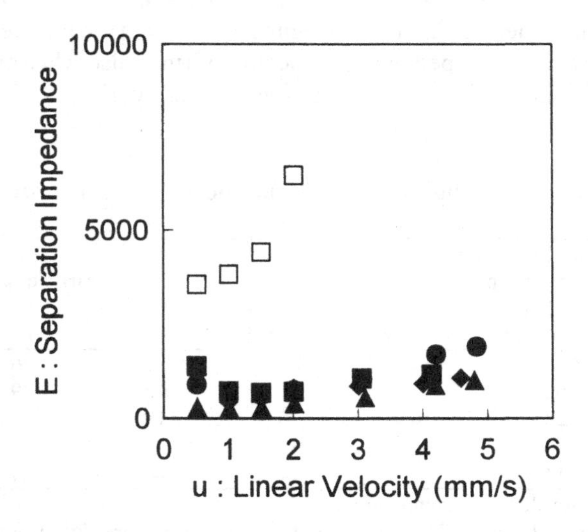

Fig. 8 Dependence of separation impedance, E, calculated from the data in Fig. 7 on mobile phase velocity, u. Symbols are the same as in Fig. 7.

Advantages of Rod Column

The chromatographic evaluation results described above show that the rod columns can be used as highly efficient alternatives of conventional particle-packed columns. The thin gel skeleton contributes to weaken the mobile phase velocity dependence of plate height, whereas the high through-pore volume does to lower the column pressure in rod columns. These features lead to a drastic reduction of separation impedance especially in the higher mobile phase velocity conditions. The physico-chemical properties of mesopores are almost the same in rods and particle-packed columns, which promises the applicability of rod columns to all the fields where silica-based particle-packed columns have been utilized. Moreover, taking advantage of the very flat *H-u* curve, a new separation mode called "flow gradient" has been developed[18]. In this method, similarly to the conventional solvent gradient, a stepwise increase of mobile phase velocity effectively accelerates the elution of a solute with longer retention time without deteriorating the peak shape. As a practical example, by only applying a higher constant flow rate and a conventional solvent gradient, the separation of mixture of polypeptides can be accelerated from 40min by a conventional method to 5min with a rod column[19].

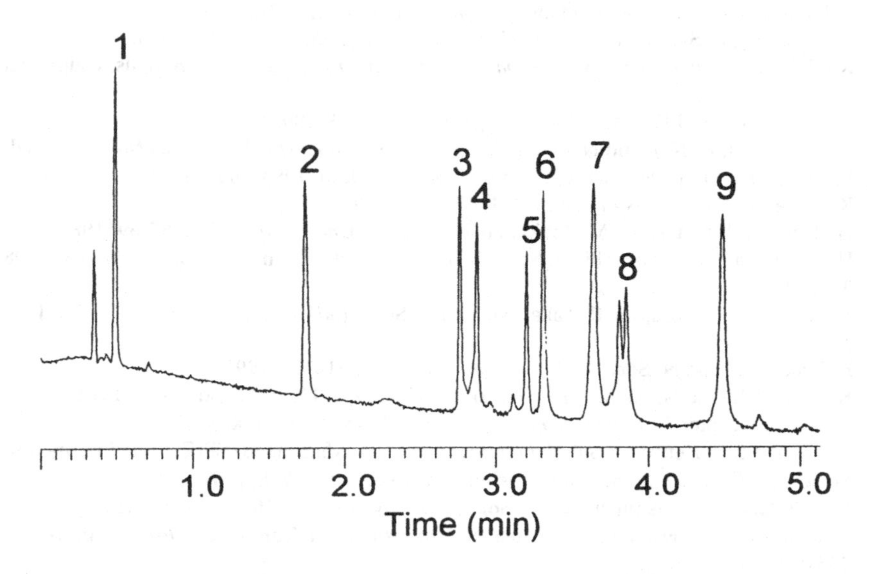

Fig. 9 Elution of polypeptides [1: glycyltylocine; 2: leucine-enkephalin; 3: insulin, 4: cytochrom c; 5: lysotyme; 6: transferrin; 7: BSA; 8: b-lactogloblin; 9: ovalbumin] on silica rod column. Linear gradient were from 5 to 60% acetonitrile in the presence of TFA at 4mms^{-1} of mobile phase velocity. Gradient time: 5min.

CONCLUSION

The monolithic silica gel column for HPLC has been developed to exhibit superior analytical performance to that of conventional ones especially at higher mobile phase velocity. The column microstructure can be designed far more arbitrarily in the rod columns. Thinner gel skeleton and higher through-pore volume in the well-defined structure both contribute to enhance the analytical performance of the rod columns under higher mobile phase velocity. Appropriate combinations of column structure, chemical modification and column geometry will open a new phase of high speed HPLC with monolithic columns.

ACKNOWLEDGEMENT

This work was supported in part by Grant-in-Aid for Scientific Research (Nos. 09554050, 09555190 and 09750751) from the Ministry of Education, Science, Sports and Culture, Japan.

REFERENCES

[1] G. Guiochon, in Cs. Horvath (Editor), *High Performance Liquid Chromatography - Advances and Perspectives*, Vol. 2, Academic Press, New York, 1980, pp. 1-56.
[2] S. Hjerten, J.-L. Liao and R. Zhang, *J. Chromatogr.*, **473**, 273-275, 1989.
[3] Q.-C. Wang, F. Svec and J.M.J. Frechet, *Anal. Chem.*, **65**, 2243-2248, 1993.
[4] R.P.W. Scott, *Liquid Chromatography Column Theory*, John Wiley & Sons, Chichester, 1992.
[5] P.A. Bristow and J.H. Knox, *Chromatographia*, **10**, 279-289, 1977.
[6] K. Nakanishi, H. Kaji and N. Soga, *Ceramic Transactions* vol. **31** "Porous Materials", eds. K. Ishizaki et al., (The American Ceramic Society, Ohio, 1993), pp. 51-60.
[7] K. Nakanishi, *J. Porous Mater.*, vol. **4**, 67-112, 1997.
[8] N. Tanaka, H. Kinoshita, M. Araki and T. Tsuda, *J. Chromatogr.*, **332**, 57-69, 1985.
[9] H. Minakuchi, K. Nakanishi, N. Soga, N. Ishizuka and N. Tanaka, *Anal. Chem.*, **68**, 3498-3501, 1996.
[10] K. Nakanishi, H. Komura, R. Takahashi and N. Soga, *Bull. Chem. Soc. Jpn.*, **67**, 1327-1335, 1994.
[11] K. Nakanishi and N. Soga, *J. Am. Ceram. Soc.*, **74**, 2518-30, 1991.
[12] K. Nakanishi and N. Soga, *J. Non-Cryst. Solids*, **139**, 1-13, 1992 and 14-24, 1992.
[13] R.K. Iler, Chapter 5 in *The Chemistry of Silica*, Wiley, New York, 1979.
[14] S. Liu and L.L. Hench, in *Sol-Gel Optics II*, ed. J.D. Mackenzie (SPIE - The International Society for Optical Engineering, Washington, 1992), vol. **1758**, p. 14.
[15] R. Takahashi, K. Nakanishi and N. Soga, *Faraday Discuss.*, **101**, 249-263, 1995.
[16] H. Minakuchi, K. Nakanishi, N. Soga, N. Ishizuka and N. Tanaka, *J. Chromatogr. A*, **762**, 135-146, 1997.
[17] H. Minakuchi, K. Nakanishi, N. Soga, N. Ishizuka and N. Tanaka, *J. Chromatogr. A*, 797, 121-131, 1998.
[18] H. Minakuchi, Doctoral Thesis, Kyoto University, Kyoto, Japan, 1997.
[19] K. Cabrera, G. Wieland, D. Lubda, K. Nakanishi, N. Soga, H. Minakuchi and K.K. Unger, *Trends in Analytical Chemistry*, **23**, 50-53, 1998.

TRANSPLANTATION THERAPIES

Edward J. A. Pope
Solgene Therapeutics, LLC (MATECH)
31304 Via Colinas, Suite 102
Westlake Village, CA 91362-3901
matech@sure.net

ABSTRACT

Solgene Therapeutics, LLC is developing a new biotechnology drug delivery system through the application of advanced sol-gel processing technology that enables the encapsulation and transplantation of live therapeutic protein secreting tissue cells without the need for immunosuppression. Solgene's patented sol-gel silica microbeads are porous enough to allow the host (patient) and implanted cells to exchange nutrients, waste products, and secreted therapeutic proteins, but exclude lymphocytes and antibodies that the host's immune system uses to attack foreign cells. This "platform" technology enables the delivery of varied biotechnology drugs, including proteins and growth hormones, secreted by genetically engineered cell lines. In initial research in the area of diabetes, encapsulated, healthy pancreatic islets were transplanted into diabetic mice, resulting in the elimination of all hematological and urological manifestations of the disease for over four months. Further research on genetically engineered cells is in progress.

INTRODUCTION

Cell therapy has emerged in recent years as an extremely promising method of treating a wide variety of disease states, including central nervous system disorders, diabetes, hormonal imbalances, liver diseases, cartilage and bone diseases, and cardiovascular disease [1]. With the exception of autologous cell therapies, in which the patient's own cells are harvested and retransplanted, as in haematopoietic cell transplantation during chemotherapy or radiation therapy, immunologic rejection poses the single greatest challenge to cell-based

To the extent authorized under the laws of the United States of America, all copyright interests in this publication are the property of The American Ceramic Society. Any duplication, reproduction, or republication of this publication or any part thereof, without the express written consent of The American Ceramic Society or fee paid to the Copyright Clearance Center, is prohibited.

therapeutics. The absence of a universal donor cell line for humans requires that either immunosuppression of the host or immunoisolation of the implanted cells be employed. *Allografts*, human to human transplantations, can be performed with immunosuppressive drug therapy, provided there is a good tissue type match. For whole organ transplants, such as kidney, heart, lung, and liver transplants, the risk-to-benefit ratio justifies immunosuppressive therapy. With immunosuppression comes a heightened risk of infections. While immunosuppression is theoretically an option for cell therapy, the risks generally outweigh the potential benefits. Therefore, cell encapsulation is the most practical approach. *Xenografts*, however, in which cells are transplanted across species, currently require isolation from the immune system period.

Cell therapy also provides a potentially improved method of biotech drug delivery. Because of their inherent instability and sensitivity to low pHs, biotech-derived proteins often possess poor oral bioavailability. Amgen's three drugs, neupogen, epogen, and infergen, for example, all must be administered intravenously. This leads to "surges" in dosage which heightens the incidence of side-effects. Moreover, the short lifetime of many therapeutically useful proteins precludes alternative long-term delivery methods, such as reservoir-based controlled release agents. The quest for better delivery methods includes: 1) mini electronic pumps; 2) inhalation delivery systems; 3) pateches and time-release polymers; 4) carrier chemicals to assist oral bioavailability; 5) whole organ transplants, and; 6) live-cell encapsulation.

Through the encapsulation and transplantation of genetically engineered cells, the biotech "factory" can be transplanted into the patient, thereby manufacturing the drug *in vivo* for the duration of therapy (see figure 1). The duration of therapy can be regulated by the timed breakdown of the encapsulating agent thereby exposing the transplanted cells to the immune system, by the preprogrammed death of the cell (apoptosis), and by external chemical triggers to cell death. Thus, biodegradable silica gel encapsulation represents a potential drug delivery platform for delivering hormones, biotech drugs, insulin, and detoxifying agents. It can also be employed for livestock and animal health applications.

The idea of encapsulating living tissue and transplanting it has been around a number of years. One of the first pioneers is Anthony Sun and co-workers [2]. Several companies have been formed to exploit this concept, particularly with regard to diabetes. All of these efforts, both in academia and the private sector, are based on *organic* encapsulation materials, primarily alginate gels derived from seaweed. While researchers in this area have experienced qualified success in demonstrating the concept of live cell encapsulation and transplantation, they have also encountered a number of adverse effects, including fibrosis (foreign body response of the immune system) around their capsules, pore blockage, and the (supposed) inability to deliver biomolecules significantly larger than insulin (6,000 kDa). We believe that *inorganic* encapsulation materials, such as silica gel, promise better biocompatability, larger total pore volumes, tailored pore diameters, a duration of therapy that can be regulated, and superb immunoisolation.

Solgene's current projected product development pipeline includes drugs for animal health and livestock (such as porcine somatatropin), human "niche" therapuetics (such as osteoinduction, biotech drugs, and antiangiogenic factors), and human chronic conditions (such as diabetes). Our research benefits from relationships with the Cornell University Medical School (Hospital for Special Surgery), the Department of Veterans Affairs (West Los Angeles VA Medical Center), and the University of California, Los Angeles (School of Engineering & Applied Science). We currently have a rapidly growing patent portfolio [3,4.5].

SILICA GEL ENCAPSULATION

The chemical routes to processing silica gel at temperatures near ambient by the sol-gel process is well understood [6]. Hydrolysis of tetraalkoxysilanes occurs most rapidly at low pH while polycondensation is most rapid at high pH. The knowledge of this led to the "Sandia Two-Step", in which TEOS hydrolyzed at low pH is then rapidly polymerized by the addition of base[6]. The gels form by a fractal process of cluster growth followed by cluster-cluster aggregation [7]. In order to encapsulate living cells, a substantially modified method of Brinker was used. The microstructure of the resulting gels reveals an aggregate of approximately 8.5 nm clusters with pores of approximately the same order of magnitude (see figure 2). The preferred geometry for cell encapsulation is small microspheres. Microspheres have been made by both *emulsion* and by *drop tower* in a biologically non-toxic oil for which the gel forming solution is immiscible. Both emulsion and drop tower derived silica gel show average pore diameters (in the wet hydrogel state) of between 10.5 and 16.1 nm (table 1). Known cell pathogenic agents, from antibodies to protozoa, range in size between 50.0 nm to over 10.0 microns (see figure 3).

In encapsulating living cells in silica gel, several key factors need to be addressed, including pH, temperature, salinity, and atmosphere (*aerobic* vs. *anaerobic*). Our earliest attempts at cell encapsulation involved the common fungi *S. cerevisiae*, in which spores of the fungi were encapsulated in silica gel and showed bioactivity even after one year held at 5 degrees C [8]. Dubbed by *Business Week* "A Ceramic Biosphere for Benign Bugs", our early research into fungi pointed the way to ever more complex and challenging research in drug delivery and diabetes [9]. Since then, numerous fungi and bacteria, as well as a wide range of mammalian tissue have been successfully encapsulated showing long-term viability and potential therapeutic value[10-12].

DIABETES

One of the most technologically challenging disease states to treat effectively is *diabetes mellitus*. In type I, insulin-dependent diabetes, the beta cells of the pancreas secrete insufficient insulin to support metabolic uptake of glucose, resulting in hyperglycemia. While normal blood glucose levels in humans average about 70 mg/dL, untreated diabetes can result in blood glucose levels as high as 500 mg/dL, resulting in appearance of long-term damage such as blindness, hypertension, coronary artery disease, renal failure, neuropathy, and the amputation of extremities. The treatment, since the early 1920s, has been daily injection of insulin. Unfortunately, poor dosing can result in hypoglycemia, in which blood glucose levels fall, which can lead to loss of consciousness, coma, and death. It is estimated that 4 - 13 percent of type I diabetics die from hypoglycemic events. The reason insulin and glucose control is so difficult is that the beta cell of the pancreatic islets of Langerhans incorporate a sophisticated biofeedback within the cells to continuously monitor blood glucose levels, glucose metabolization, and ATP synthesis rates in the mitochondria, to constantly adjust the level of insulin (and type) excreted by the cells.

No artificial system has yet been developed that accurately mimics the sophistication of this natural system. Therefore, about 20 years ago, the concept of treating type I diabetics with encapsulated, healthy pancreatic islets from donor sources was proposed [2]. Proposed donor sources include cadavers, fetal tissue, and porcine xenografts. All three primary tissue sources have their own drawbacks, most notably the transfer of infectious agents (HIV, hepatitis, etc. for human tissue sources) and the creation of new xenozoonoses (from cross-species transplantations). More recently, concern has arisen over the transfer of prions, such as that implicated in bovine spongiform encephalopathy. Despite these and other concerns, the field of encapsulated islet transplantation to treat diabetes holds great promise.

In order to investigate whether silica gel encapsulation methods may provide an attractive alternative to the more conventional alginate materials which currently predominate, an *in vitro* and *in vivo* experimental protocol was established. Many of the specific details have been published previously [11,12]. The general processing and experimental steps are: harvesting donor pancreata; enzymatic digestion to isolate the islets from acinar tissue; islet viability testing; silica gel encapsulation; capsule viability testing; peritoneal or subcutaneous implantation, and; *in vivo* performance. Donor tissue was harvested from C57ob/ob mice, followed by enzymatic digestion in collagenase/dispase solution. Upon verification of islet viability, islet suspensions were encapsulated as described previously [11,12] and cultured in growth media (bovine growth sera) and subjected to further viability tests (FSR). The prefered capsule size is < 500 microns to permit ample diffusion of metabolites to all beta cells in the capsules (see figure 4).

Viable capsules were transplanted into male NOD mice which were rendered diabetic through the administration of streptozotocin. All transplant recipients had a blood glucose level > 400 mg/dL prior to implantation. Our earliest success was dubbed "supermouse" with a complete elimination of measured hyperglycemia for over two months. It was determined that the duration of treatment was limited by the time required for the capsule to biodegrade. Subsequent tests with an improved silica gel formulation resulted in similar success in excess of 4.5 months. In a data set of over thirty mice, divided into four categories: 1) streptozotocin-treated; 2) streptozotocin-treated plus placebo capsules; 3) streptozotocin-treated plus large (3.0 mm) drop tower capsules with islets, and; 4) streptozotocin-treated plus small (0.3 mm) emulsion capsules with islets, the difference in morbidity rate was extremely dramatic. While the strep-treated mice, in which no intervention was attempted, lived an average of only 9 days, mice receiving the emulsion processed islet transplants survived over 100 days, with the longest survivor lasting 154 days!

GENETICALLY ENGINEERED AND OTHER CELLS

While diabetes posses the greatest therapeutic challenge due to the long-term nature of the therapy, many other shorter term therapies may be easily obtainable. Quarterly or bi-annual injections of <500 micron silica beads encapsulating a variety of therapeutic protein secreting cells could provide an attractive alternative to conventional daily injections or numerous orally ingested pills several times daily (for example, AIDS patients). A number of other possible applications are currently being explored.

Numerous diseases of the central nervous system exist due to the inability or low ability to produce key neurotransmitters. One example is Parkinson's disease, in which the lack of sufficient dopamine leads to severe motor difficulties. A brief attempt was made (with Pat Tresco, University of Utah, Salt Lake City) to encapsulate the dopamine-secreting PC-12 cells of rats. Cell viability was confirmed by vital dye staining techniques. Further work is warranted.

In an attempt to demonstrate the ability to encapsulate genetically-engineered cells, the classic biotech system of incorporating the lac-z gene (which signals the production of beta-galactosidase) into mouse mesenchymal stem cells using an adenoviral vector was selected. Working with J. Lane and A. Tomin (Cornell University Medical School), lac-z transfected stem cells were encapsulated and demonstrated to be viable *in vitro* after 10 days (table 2). Cells were exposed to an indicator dye for the presence of beta-galactosidase. The lac-z transfected cells turned blue, indicating the presence of the protein, while the encapsulated control cells (without the gene) remained white.

A more ambitious study with Cornell has just begun involving 95 rats testing for the ability of silica encapsulated osteosarcoma cells which produce bone morphogenic proteins to stimulate bone formation. In addition to osteoinduction, immunochemistry, toxicology, and histology studies will also be conducted as part of the five month joint effort.

SUMMARY

A new and promising delivery platform, based upon sol-gel silica encapsulation of living cells, has been demonstrated. Pore size is of sufficient size to simultaneously allow large biomolecules to pass while excluding the larger antibodies of the immune system. Preliminary diabetes studies on 30 mice demonstrate the potential for long-term therapy for diabetes. Genetically engineered cells designed to secrete biomolecules have been demonstrated. Current research on bone regeneration, involving a 95 rat study, hold great promise. This technology offers competative advantages for the delivery of varied biotech drugs.

ACKNOWLEDGEMENTS

This work would not be possible without the collaboration and assistance of a number of very talented individuals: C. M. Peterson and K. P. Peterson (Diabetes), A. Tomin and J. Lane (lac-z transfection, osteoinduction and bone growth), P. Tresco (PC-12 cells), and A. Almazan (silica gel formation). The support of NIH-NIDDK (partial support of diabetes research) is greatly appreciated.

REFERENCES

1. Fred H. Cage, "Cell Therapy", *Nature*, **392 (supp)**, 18-24, (30 April 1998).
2. A. M. Sun, et al., "Microencapsulation of Living Cells as Bioartificial Organs", in *Artificial Organs*, edited by J. D. Andrade, et al., VCH Publishers, New York, 1987.
3. E. J. A. Pope, "Encapsulation of Living Tissue Cells in an Organosilicon Gel", U. S. Patent 5,693, 513, December 2, 1997.
4. E. J. A. Pope, "Sol-Gel Encapsulation of Living Animal Tissue Cells and Living Micro-organisms", U. S. Patent 5,739,020 (April 14, 1998).
5. several further U. S. and foreign patents pending.
6. C. J. Brinker and G. W. Scherer, *Sol-Gel Science*, Academic Press, New York, 1990.
7. E. J. A. Pope and J. D. Mackenzie, "Theoretical Modelling of the Structural Evolution of Gels", in *Journal of Non-Crystalline Solids*, **101**, 198-212 (1988).
8. E. J. A. Pope, "Gel Encapsulated Microorganisms: S. cerevisiae-Silica Gel Biocomposites", *Journal of Sol-Gel Science and Technology*, **4**, 225-229 (1995).
9. E. T. Smith, "A Ceramic Biosphere for Benign Bugs", *Business Week* p. 105 (November 7, 1994).
10. E. J. A. Pope, et al., "Living Ceramics"; pp. 33-49 in *Sol-Gel Science and Technology*, Edited by E. J. A. Pope, S. Sakka, and L. C. Klein, American Ceramic Society, Ceramic Transaction volume 55, Westerville, 1995.
11. E. J. A. Pope, K. Braun, and C. M. Peterson, "Bioartificial Organs I: Silica Gel Encapsulated Pancreatic Islets for the Treatment of Diabetes Mellitus", *Journal of Sol-Gel Science and Technology*, **8**, 635-639 (1997).
12. K. P. Peterson, C. M. Peterson, and E. J. A. Pope, "Silica Sol-Gel Encapsulation of Pancreatic Islets", *Proceedings of the Society of Experimental Biology and Medicine*, accepted.

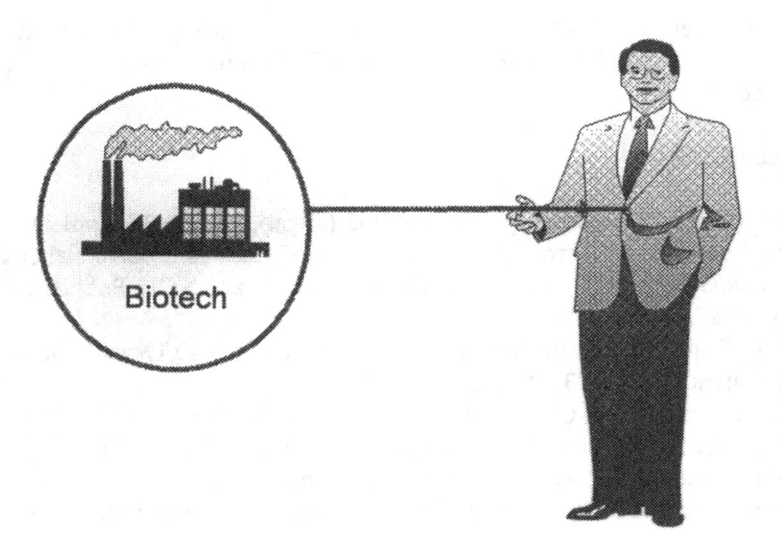

Figure 1: Through silica gel microencapsulation of genetically engineered cells, Solgene Therapeutics hopes to take the biotech factory and transplant it into the patient to continuously manufacture therapeutics *in vivo*.

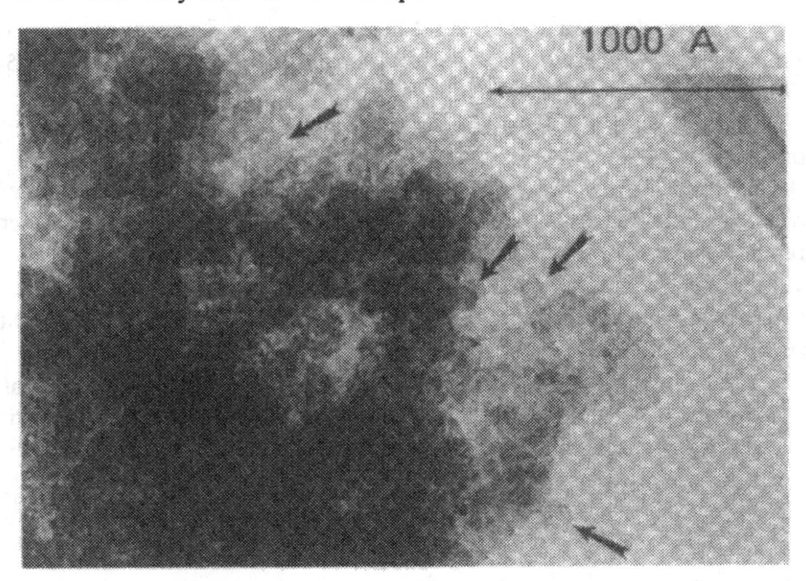

Figure 2: T. E. M. of silica gel structure illustrating aggregates of approximately 8.0 nm clusters

Table 1: Comparison of physical properties of drop tower and emulsion derived silica gel microencapsulants.

	Drop Tower	Emulsion
Average Diam. (mm)	3	0.3
Surface Area (mm2/gm)	817	996
Pore Volume (cc/gm)	3.3	2.62
Ave. Pore Diam. (nm)	16.1	10.5
Peak Pore Diam. (nm)	8.9	6.5

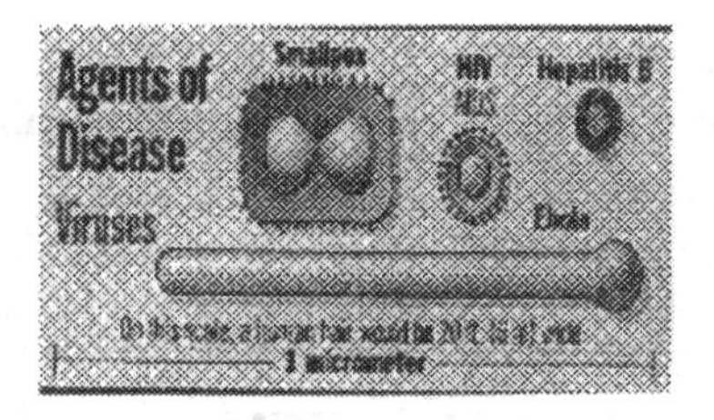

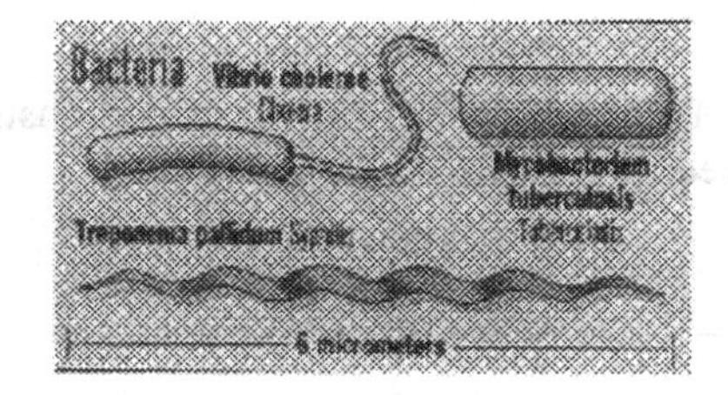

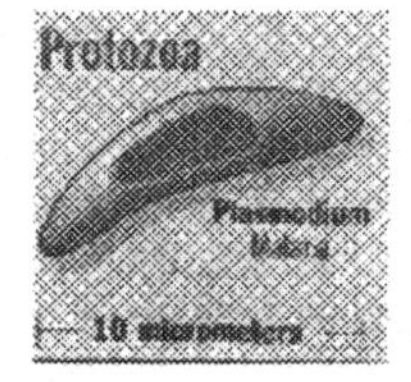

Gel Pore Size ~100A

Figure 3: Comparison of characteristic sizes of various infectious agents.

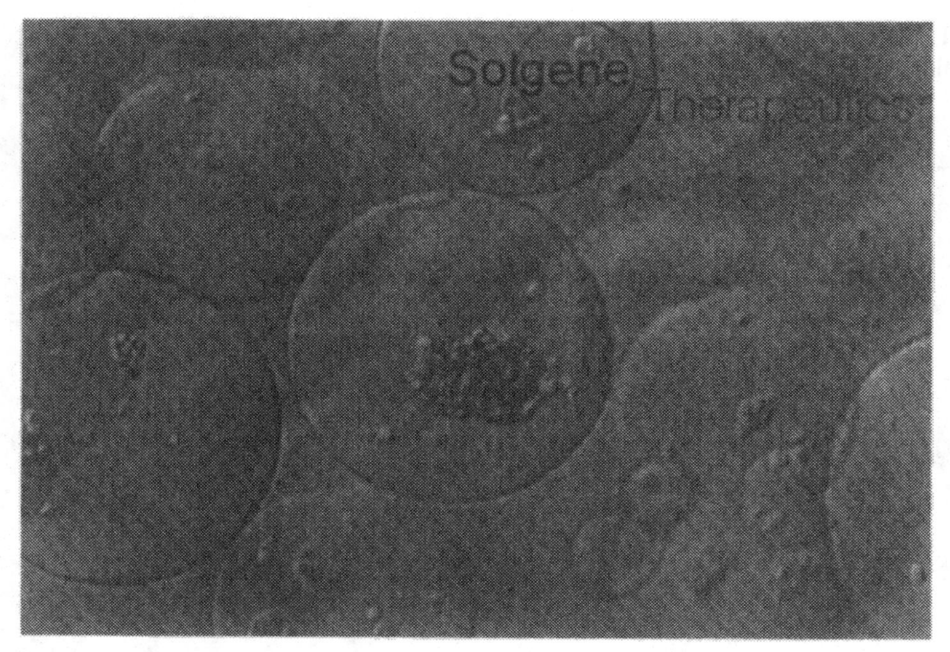

Figure 4: Photomicrograph of silica gel encapsulated porcine pancreatic islets in ~500 micron diameter bead.

Table 2: Example system chosen to demonstrate ability to encapsulate genetically engineered cells.

CLASS	TEST SYSTEM
Gene	lac-z
Vector	adenovirus
Host	mesenchymal stem cells

ULTRA LIGHT CERAMIC FOAMS

G.S. Grader, Y. de Hazan and G.E. Shter
Chemical Engineering Department, Technion, Haifa 32000, ISRAEL

ABSTRACT

In this paper we report a novel process for the production of ultralight cellular ceramics. The foams are produced by the heat treatment of crystals of the $AlCl_3(Pr^i_2O)$ complex. The foams consist of an arrangement of closed cells, 100-300 μm in diameter, having cell walls approximately 1-2 μm thick. Surface area measurements show that the cell walls contain nanometer-sized pores. The cell morphology is retained at 1500^{0}C with only a small shrinkage in the foam dimensions. The void fraction in the foams can be as high as 99% - the highest porosity reported in cellular ceramics. The mechanism of foam formation in this system is described. The key factors in this process are the simultaneous gelation and evolution of a volatile nonhydrolytic reaction product, which is triggered by a phase separation in solution.

INTRODUCTION

Ceramic foams have been used extensively in filtration of molten metals, in catalyst supports, and other high temperature applications. Traditionally the production is based on infiltration of a pre-existing organic skeleton (i.e. polyurethane foam)[1,2] and the void fraction is in the 70-90% range. More modern non-impregnation methods include polymerization of an emulsion,[3] and the introduction of physical and chemical foaming (blowing) agents. The latter techniques require that the agent be activated when the viscosity is sufficiently high to prevent the collapse of the foam structure. These approaches include the work by Fujiu et al.[4] and by Wu and Messing[5], where freon was introduced into a sol at low temperature and used as a physical blowing agent upon heating the system above the freon boiling point. Careful timing had to take place to ensure that the boiling takes place during gelation, where the viscosity was high enough. Typical porosity was 85%. Alternatively, a chemical blowing agent can be used or the foam structure stabilized by the polymerization of organic monomers incorporated into a suspension[6].

We have recently described a novel method to produce ceramic foams using a chemical blowing agent that is a reaction product of a nonhydrolytic sol-gel process.[7-9] This method is based on the non-hydrolytic reactions[10-13] that lead to a partially condensed metal oxide and a potentially volatile alkyl halide. This generic foaming method was demonstrated with aluminum chloride and isopropyl ether.[7-9] The method is simple and self regulating since the blowing agent is activated naturally and simultaneously with the gelation process. The method is distinguished by a phase separation that takes place in a solution containing partially condensed Al-O-Al species and isopropyl chloride product. Upon separation into solvent rich regions and polymer rich regions the solvent starts boiling, while the aluminum species undergo gelation. It is the combination

To the extent authorized under the laws of the United States of America, all copyright interests in this publication are the property of The American Ceramic Society. Any duplication, reproduction, or republication of this publication or any part thereof, without the express written consent of The American Ceramic Society or fee paid to the Copyright Clearance Center, is prohibited.

of phase separation, chemical blowing agent and simultaneous gelation that distinguishes this foaming scheme from the other ceramic foaming processes. The net result of this scheme is the formation of highly porous foams with cell walls in the 1-2 μm range and cell size of 50-300 μm . The void fraction in these foams is in the 94-99% range - the highest reported porosity in a cellular ceramic and second in density only to silica aerogels produced by super critical drying (SCD) procedures.

The main purpose of this write-up is to describe the principles of the new foaming procedure and outline the effect of some processing parameters on the foam microstructure.

EXPERIMENTAL PROCEDURES

Preparation of nonhydrolytic alumina sols was presented elsewhere.[10-13] Crystals of the $AlCl_3(Pr^i_2O)$ complex were prepared from solutions with Al/CH_2Cl_2 molar ratio of 1:2.4. The $AlCl_3/Pr^i_2O$ ratio of 2:3 was used. The detailed preparation scheme was reported previously.[7,8] The dry crystals were kept in sealed vials filled with argon at 5^0C until the foaming experiments. Due to their sensitivity to moisture, all manipulations of the crystals were carried out in argon or in vacuum. Heat treatment of the foams was performed in argon or air up to a maximum temperature of 1500^0C.

To elucidate the foaming mechanism, 1 gram of crystals were added to a 200 ml sealed glass reactor. The reactor was then placed in an oil bath at about 70^0C, and examined continuously using a video camera.

DSC measurements of the crystals and sols were conducted using a Perkin-Elmer DSC-7 apparatus. Heating rate was 5^0C/min. The gases evolved during the decomposition of the crystals and after the formation of the foam were isolated and analyzed by NMR, Bruker 200 Herz, and by Perkin-Elmer Autosystem XL gas chromatograph.

The specific surface area of the foams was determined by single point BET method via nitrogen adsorption/desorption (Micromeritics, Flosorb II). DTA/TGA measurements were conducted in a Setaram TG-92 unit in air. Heating rate was 5^0C/min. The morphology was observed using a SEM (Jeol 5400).

RESULTS AND DISCUSSION

As shown elsewhere,[7-9] the starting point for the nonhydrolytic ceramic foams are needle-like crystals of the $AlCl_3/Pr^i_2O$ complex. The existence of $AlCl_3$/ether complexes is well documented in the literature[10-11]. The ratio of $AlCl_3$/ether in the complexes is normally 1:1 but can be also 1:2 or 2:1 depending on the nature of the ether. The TGA/DTA plot in Fig. 1 shows the decomposition of the crystals upon heating in air (5^0C/min). From the overall weight loss of 78%, the ratio of $AlCl_3$/ether in the complexes is calculated to be 1:1. The initial endotherm up to ~150^0C is due to the evolution of volatile species, where most of the weight is lost. The shallow and wide exotherm in the 200-600^0C range is due to oxidation of residual carbon in the material. Finally, the small exothermic peak at ~850^0C is due to the primary crystallization of η-Al_2O_3.[13] Previous work had shown that nonhydrolytic alumina xerogels crystallize in the 840-820^{0}C range depending on the aging conditions.[12,13] The higher crystallization temperature observed here is consistent with the short gelling and lack of aging period in the foams.

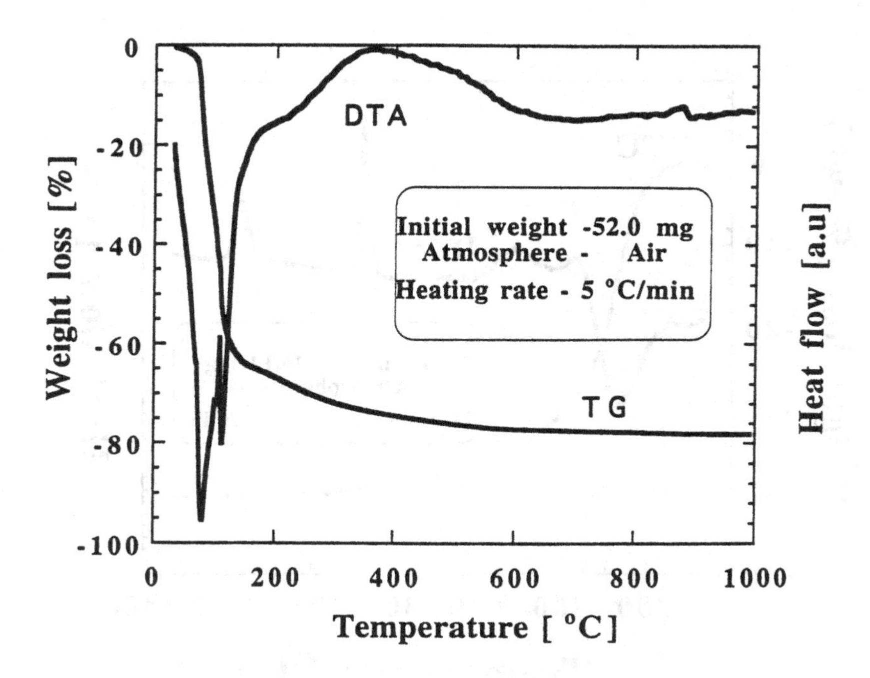

Fig. 1. DTA/TGA curves of crystals of the $AlCl_3(Pr^i{}_2O)$ complex, heated in air at 5^{0}C/min.

The relative size of the crystallization and the oxidation peaks in Fig. 1 should be compared with their counterparts during the heat treatment of nonhydrolytic alumina xerogels, shown in Fig. 2. The larger relative size of the crystallization peak in Fig. 2 stems from the fact that the mass of the xerogel rests on the bottom of the TGA crucible, in good contact with the DTA thermocouple.

On the other hand, the crystals examined in Fig. 1 foamed within the crucible during the heat process. Therefore the bulk of the mass in the crucible was away from the DTA thermocouple at the bottom - giving rise to a poor heat flow signal during the crystallization of η-Al_2O_3. The second exothermic peak in Fig. 2 at ~1100C, displays the polymorph phase transition into α-Al_2O_3.

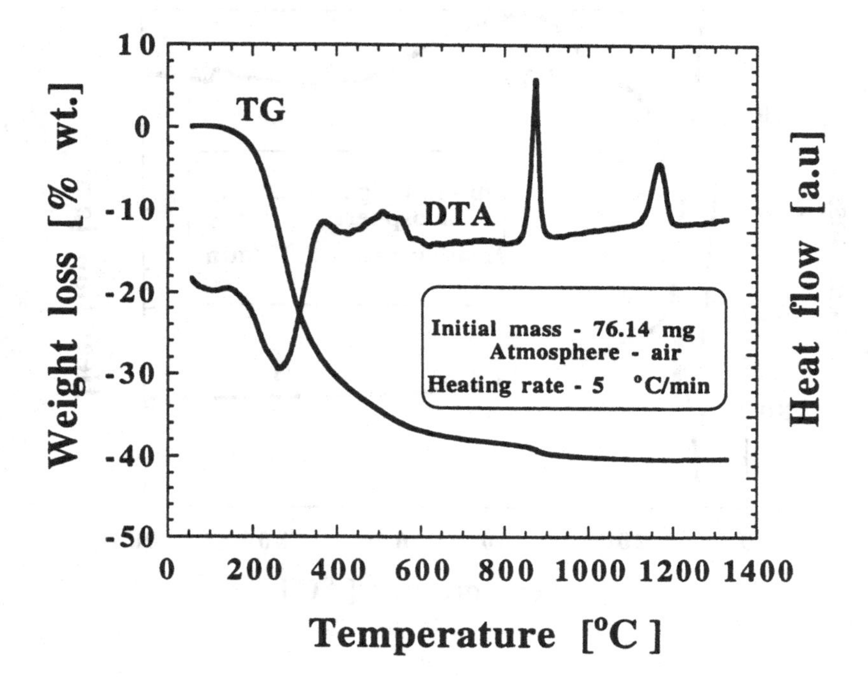

Fig. 2. DTA/TGA curves of nonhydrolytic alumina xerogels, heated in air at 5^0C/min.

The stages of the foaming process at 70^0C in a sealed 200 ml reactor are shown in Fig.3a-e. Fig.3a shows the dry crystals before the heating stage. Upon heating of the crystals to temperatures of about 70^0C, a liquid phase appears and dissolves the crystals, the crystals become transparent, and some gas is evolved. A partial dissolution of the crystals with a liquid phase is shown in Fig. 3b. Subsequently, a homogeneous liquid phase is obtained as evident in Fig.3c after

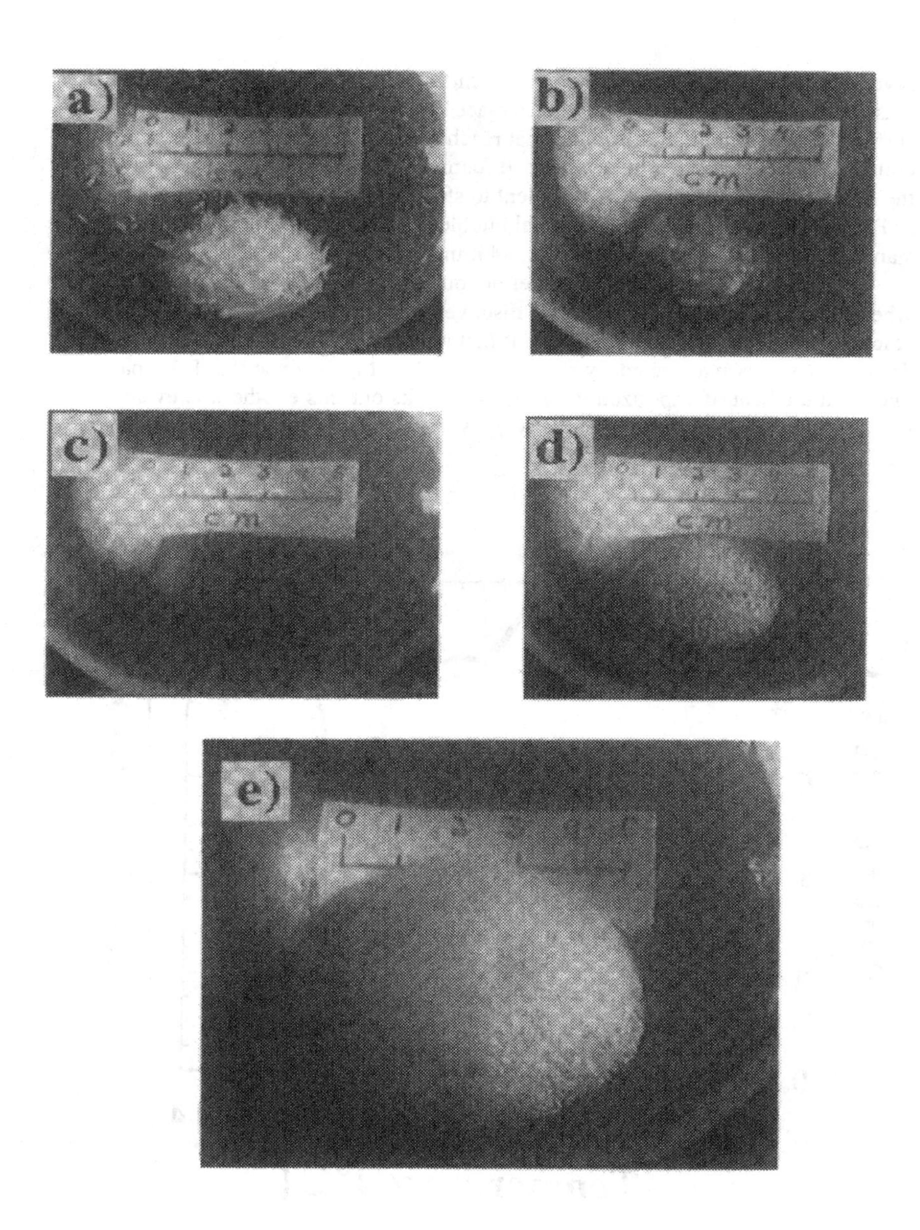

Fig. 3. Stages in the formation of foam from crystals at 70ºC: a) initial crystals; b) partial crystal dissolution; c) full dissolution; d) initial foaming; e) final foam.

several minutes. Depending on the liquid temperature, after few more minutes a massive nucleation of gas bubbles takes place. The gas bubbles rise to the liquid surface as shown in Fig.3d. Bubbles that reach a critical size on the surface collapse, however, further nucleation and foaming continues until the viscosity of the liquid reaches a level that is sufficient to stabilize the foam structure presented in Fig.3e. The total time from the initial bubble nucleation until the full stabilization of the foam structure was 2-4 minutes at 70^0C.

In a separate experiment a thermocouple was placed in the glass vessel where the foaming took place. It was discovered that the temperature rose in the reactor during the foaming, pointing out that an exothermic event is taking place. This exotherm is undetected by the DTA signal in Fig. 1 since the TGA pan is open and the heat of vaporization of the gas masks out this exothermic event. To detect this exothermic peak quantitatively, some crystals were heated in a closed DSC pan.

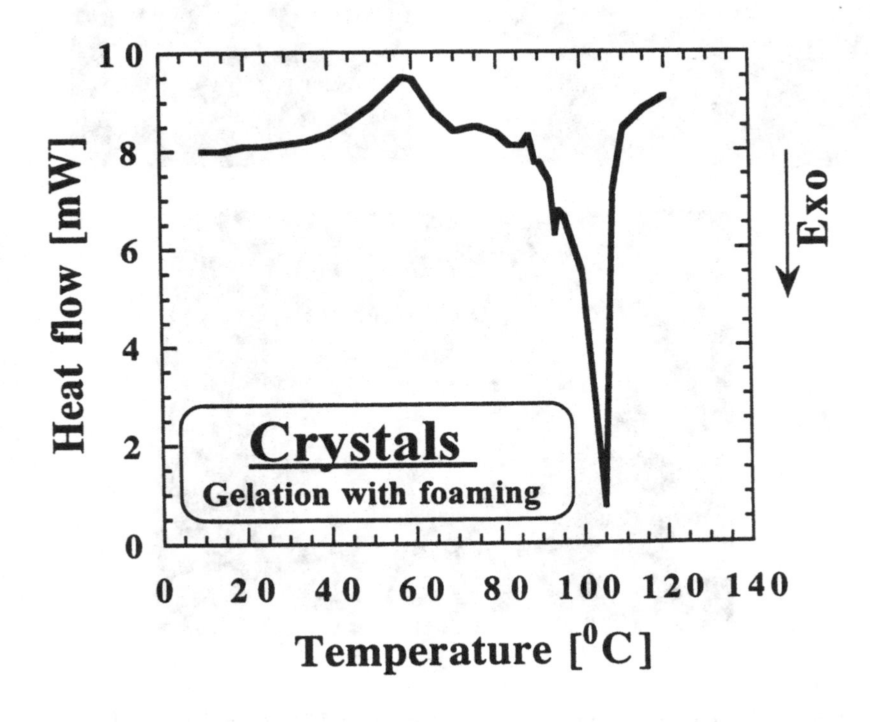

Fig. 4. DSC of crystals of the $AlCl_3(Pr^i_2O)$ complex in a closed capsule.

The results are shown in Fig. 4, where the small endo-effect at 40-60^{0}C is due to slight evaporation, and the large exo-effect at 80-110^{0}C is analogous to the one seen in the bulk foaming experiment. We believe that this exotherm is due to gelation. To prove that the gelation is indeed exothermic we repeated the DSC experiment with a sol that gels upon heating but does not undergo foaming. The nonhydrolytic sol contained 3% crystals of the $AlCl_3(Pr^i_2O)$ complex dissolved in dichloromethane. Results of this experiment shown in Fig. 5 clearly demonstrate the gelation is exothermic, and therefore the gelation occurs simultaneously with the foaming sequence shown in Fig. 3.

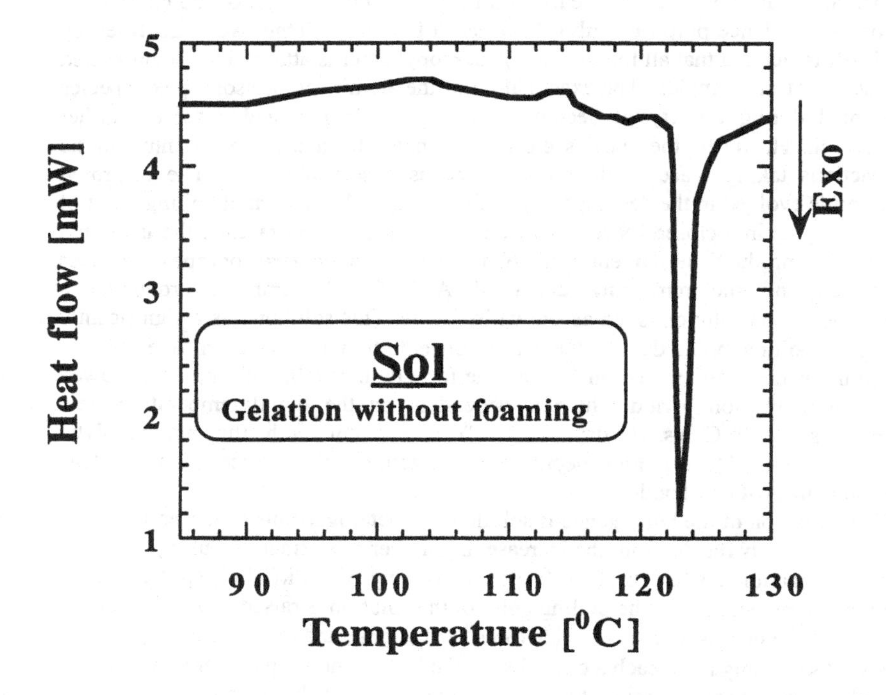

Fig. 5. DSC of nonhydrolytic alumina sol in a closed capsule.

An example of a foam treated at 1300^{0}C is shown in Fig 6. This foam has closed cells, 50-300 µm in diameter. However, foams with opened cells can be prepared as well. The cell wall thickness was found to be 1-2 µm . The void fraction of various foams was 94-99%. This is the highest porosity ever achieved in ceramic cellular structures. Such a high porosity has been reported in silica aerogels (prepared by super critical drying of gels), however their structure is fractal rather than cellular.

Fig.3a indicates that the first stage of the foaming process involves the gradual appearance of a liquid phase accelerated by heating the crystals of the $AlCl_3(Pr^i_2O)$ complex. The gas evolved in a controlled heating of the crystals to 70^0C was analyzed by means of NMR, before a liquid phase has formed. The only volatile species isolated was isopropyl chloride. This result indicates that the $AlCl_3(Pr^i_2O)$ complex is decomposed during heating with the formation of isopropoxy groups and isopropyl chloride through the following scheme:

$$Cl_3Al : O\mathrm{Pr}_2^i \xrightarrow{\Delta} Cl_2Al - O\mathrm{Pr}^i + \mathrm{Pr}^i\,Cl \qquad (1)$$

(It should be stressed that even the initial aluminum species are most likely different than those shown in eq.1, and that this representation of the complex is for mass balance purposes only). No traces of isopropyl ether were identified by NMR, indicating that all the remaining isopropyl ether is attached to the unreacted fraction of the complex. The exact nature of the aluminochloroisopropoxy species formed during the crystal decomposition is not known and deserves further research. However, the species clearly originates from the same nonhydrolytic reactions taking place in dichloromethane as a solvent.[10,11] The isopropyl chloride evolved in the decomposition of the crystals has a normal boiling point of 35.4^0C.[14] In a closed reactor however, or when liberated rapidly, the isopropyl chloride product subsequently dissolves the aluminochloroisopropoxy species formed and the remaining unreacted $AlCl_3(Pr^i_2O)$ complex, producing a homogeneous solution, such as shown in Fig.3c. This solution has a significantly higher boiling point due to the large concentration in aluminum species. No boiling is therefore evident in Fig.3c. The formation of a liquid phase is followed by an incubation period which is dependent on the liquid temperature (few minutes at 70^0C vs. 1 day at 25^0C). Consistent with the nonhydrolytic chemistry[10,11] a homogeneous polymerization takes place at this stage producing Al-O-Al bonds.

The transition of the homogeneous solution to a heterogeneous one, seen in Fig. 3d, most likely results from the increase in polymer mass fraction during polymerization. As long as the solvent and growing $AlO_xCl_y(OPr^i)_z$ species are mixed homogeneously, the boiling point of the solution is raised above the boiling point of the pure isopropyl chloride (35.4^0C at 1 atm), and its vapor pressure is low. As the polymers reach a critical size, the homogeneous polymer solution undergoes a phase separation to a polymer and solvent rich regions. At this instance, which is recognized as the foaming point, two different processes take place simultaneously in the heterogeneous mixture. In the solvent rich region, the expelled solvent is suddenly above its boiling point, or with a significant vapor pressure. Micron sized solvent bubbles nucleate within the solution, expand and rise to the surface, as a result of buoyant forces. As long as the viscosity of the polymer rich regions is low, the expansion, rise, coalescence, growth and collapse of the bubbles continues.

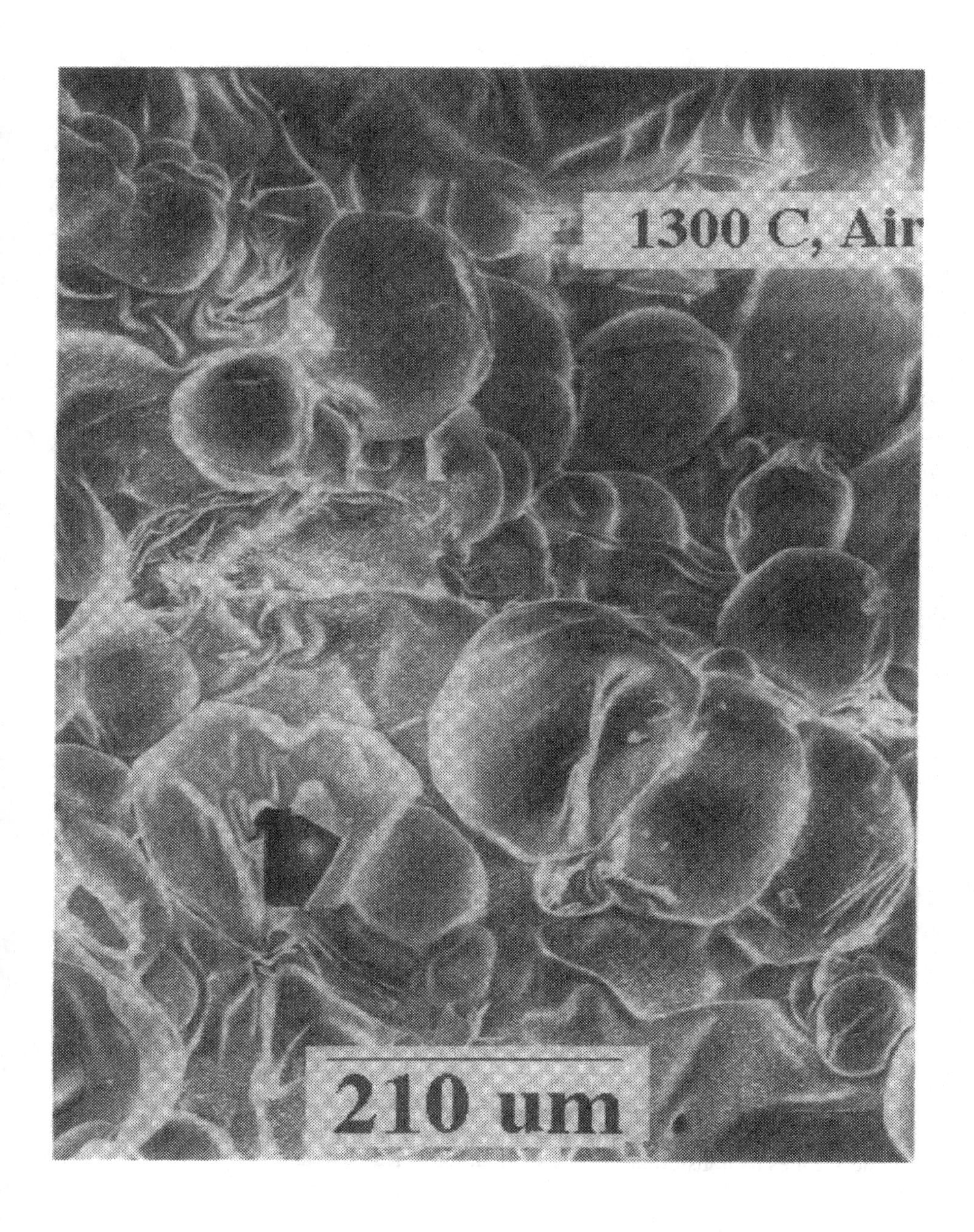

Fig. 6. Morphology of alumina foams treated at 1300^{0}C.

However, since the polymer concentration increases upon the formation of a polymer rich region, the rate of polymerization is significantly increased as well. Thus, the viscosity and surface tension of the polymer rich phase rises rapidly, resulting in much smaller, more stable gas bubbles. The foam is subsequently stabilized by gelation of the polymer phase.

Fig.7 summarizes the stages involved in the foaming mechanism from the crystals of the $AlCl_3(Pr^i{}_2O)$ complex. The process requires the crystals of the $AlCl_3(Pr^i{}_2O)$ complex as the only precursor. This foaming process is undoubtedly the simplest reported thus far with respect to starting materials and process control. The process simplicity stems from the fact that all the functions required for foam formation are embedded in only one precursor, namely the crystals of the $AlCl_3(Pr^i{}_2O)$ complex.

The decomposition of the crystals of the $AlCl_3(Pr^i{}_2O)$ complex leads to the formation of isopropyl chloride and aluminochloroisopropoxy species, both not present initially. Foam formation is only made possible due to the inherent duality of the roles played by these two products. Isopropyl chloride takes the dual role of the solvent and the foaming agent present in the solvent rich phase after phase separation. The aluminochloroisopropoxy species take the dual role of the polymerizing species giving rise to a phase separation (and additional isopropyl chloride) and the gelling/setting agents present in the polymer rich phase after phase separation. Moreover, the duality of the components, which change their role at the foaming point, where phase separation takes place, results in three simultaneous processes, namely foaming, gelling and drying. The net result is a foamed, stabilized, dried material.

SUMMARY

We report here for the first time a novel method to prepare ultra light ceramic foams, having a void fraction as high as 99%. The foams consist of an arrangement of closed or opened cells 50-300 μm in diameter with cell walls approximately 1-2 μm thick. The cell morphology is retained at 1500^0C after partial sintering which induces only a small shrinkage in the foam dimensions. The foaming mechanism from the crystals of the $AlCl_3(Pr^i{}_2O)$ complex has been elucidated. The mechanism involves two consecutive nonhydrolytic sol-gel chemical reactions and physical processes including crystal dissociation, solvation, phase separation and foaming. While other foaming mechanisms cited in the literature utilize one or more of the processes above, no analog mechanism exists in the organic, ceramic or metal foam production processes. The effectiveness of the process originates from an initial precursor which contains all the necessary components in such a way that the application of mild heating accelerates its transformation to a solid, dry, foamed material.

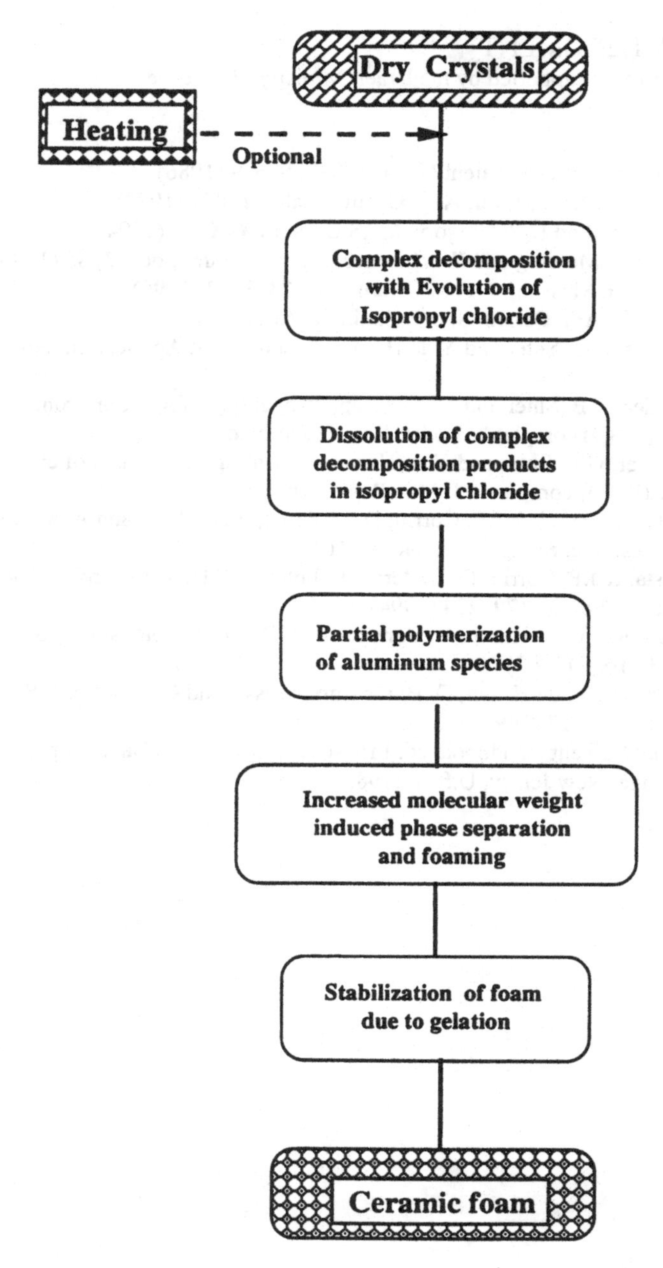

Fig. 7. Foaming stages of crystals of the $AlCl_3(Pr^i_2O)$ complex.

ACKNOWLEDGMENT

This research was supported by the Israeli Ministry of Science.

REFERENCES

1. J.W. Brockmeyer, U.S. Patent No. 4610832, Sep. 9 (1986)
2. F.F. Lange and K.T. Miller, Adv. Ceram. Mater. **2**, 827 (1987).
3. W.R. Even Jr. And D.P. Gregory, MRS Bulletin **XIX**, 29 (1994).
4. T. Fujiu, G.L. Messing and W. Huebner, J. Am. Ceram. Soc. **73**, 85 (1990).
5. M. Wu and G. Messing, J. Am. Ceram. Soc. **73**, 3497 (1990).
6. P. Sepulveda, Am. Ceram. Soc. Bulletin **76**, 61 (1997).
7. G.S. Grader, G.E. Shter and Y.de Hazan , Israeli patent Application No. 123969 (1998).
8. G.S. Grader, G.E. Shter and Y.de Hazan, "Novel foam formation from crystals of $AlCl_3(Pr^i_2O)$ complex", J. Mat. Res., (submitted).
9. G.S. Grader, G.E. Shter and Y.de Hazan, "Foaming mechanism of crystals of the $AlCl_3(Pr^i_2O)$ complex", J. Mat. Res., (submitted).
10. S. Acosta, P. Arnal, R.J.P. Corriu, D. Leclercq, P.H. Mutin and A. Vioux, Mater. Res. Soc. Symp. Proc. **346**, 43 (1994).
11. S. Acosta, R.J.P. Corriu, D. Leclercq, P. Lefevre, P.H. Mutin and A. Vioux, J. Non-Cryst. Solids. **170**, 234 (1994).
12. G. S. Grader, Y. de Hazan, Y. Cohen and D. Bravo-zhivotovskii, J. Sol-Gel. Sci.Tech. **10**, 5 (1997).
13. G. S. Grader, Y. de Hazan, D. Bravo-zhivotovskii and G.E. Shter, J. Sol-Gel., Sci.Tech. **10** , 127 (1997).
14. Industrial Solvent Handbook, ed. Ernest W. Flick, Noyes Data Corporation, Park Ridge, New Jersey, U.S.A. (1985).

FABRICATION AND CHARACTERIZATION OF SOL GEL MONOLITHS WITH LARGE MESOPORES

Kevin W. Powers
Department of Materials Science and Engineering
University of Florida, Gainesville, FL 32611-6400

Larry L. Hench
Department of Materials
Imperial College of Science, Technology and Medicine,
Prince Consort Road, London SW7 2BP

ABSTRACT

A system of producing small silica gel monoliths with uniform mesopores as large as 40nm in radius has been developed and characterized. The one step HF catalyzed process results in net shape pieces with pore volumes on the order of 1.6 cm^3/g and specific surface areas of 150 m^2/g. Pore texture characteristics can be adjusted to meet specific requirements. The porous glass is suitable for doping with a variety of organic or inorganic second phases. Aging, drying and densification characteristics have been explored and fully dense high purity silica can be produced at 1200°C under ambient atmospheres.

INTRODUCTION

There are several methods for making large pore silica gels starting from a variety of precursors. [1-4] The term "large pore" is relative but is used here to refer to any gel with a pore radius greater than 10nm. In generally accepted terminology these large pore gels with radii of up to 25 nanometers are still considered mesoporous.[5] The gels in this study have pores ranging from 10 to 40 nanometers in radius. They are produced in a single step as monoliths with very narrow pore size distributions. There is no base aging or acid etching thus reducing processing time and preserving better homogeneity and optical properties. Optical transparency in the visible is affected by the pore size due to Rayleigh scattering.

To the extent authorized under the laws of the United States of America, all copyright interests in this publication are the property of The American Ceramic Society. Any duplication, reproduction, or republication of this publication or any part thereof, without the express written consent of The American Ceramic Society or fee paid to the Copyright Clearance Center, is prohibited.

The size of the pores and large pore volume allows doping with a wide variety of materials to form chemical sensors, solution hosts and nanocomposite materials. Depending on the nature of the dopant, the resulting composite can be used in optical applications in the UV, visible and near IR regions. The large pore gels retain enough strength (relative to capillary forces) to be rehydrated after drying without the failure of the monolith. The ability to rehydrate the gels without high temperature stabilization treatments allows organic and other temperature sensitive dopants to be introduced during gelation and be retained intact in the dried gels. If pre-doping is not required and greater strength is desired the gels can be stabilized at a variety of temperatures up to 1150°C where they generally achieve full density. As stabilization temperature increases pore size remains fairly constant up through 1000°C while pore volume decreases only slightly. Structural hardness increases slightly with increasing stabilization temperature but is primarily a function of bulk density.

METHODS AND MATERIALS

The basic steps in producing silica gel monoliths include the formulation, gelation, aging, drying, stabilization and densification. Under the right conditions, doping the gels with second phases can be accomplished at any stage of the processing. The silica source used is tetramethoxy silane (TMOS), chosen because of its rapid hydrolysis under acidic conditions. The use of fluoride as a catalyst increases the condensation rate (gelation rate) thus it is convenient to have rapid and complete hydrolysis in order to better control the kinetics.[6] Consequently, the gels are prepared under acidic conditions and are condensation limited. After gelation, the aging and drying steps of the process are the second major determinant of pore size and texture.

FORMULATION

Large pore silica gels can be made by HF catalysis of tetramethoxysilane (TMOS) and water in a nitric acid/methanol solution of pH 2 - 2.5. The molar ratio (R value) of water to TMOS is 16:1. Methanol is added in equal volume proportion to water. The combination of low pH, fluoride catalysis, rapid gelation and the aging/drying processes are responsible for the large pore size. The initial pore texture is controlled by the conditions of gelation, principally the concentration of fluoride, the pH and the R ratio. Table I presents the formulation for silica gels with a variety of pore sizes. The "standard" formulation used for these studies produced an average pore radius in the vicinity of 200Å, a specific surface area of about $150m^2/g$, and a pore volume of about 1.6 cc/g. Extended periods of aging or drying, or changes in the pore liquor during aging can significantly affect the pore texture as will be discussed later. Figure 1 shows a photograph of a variety of 200Å monoliths.

Table I. Formulations for HF catalyzed TMOS derived monoliths.[7]

Pore Radius Å	H_2O (ml)	CH_3OH (ml)	TMOS (ml)	HF (3%) (ml)	HNO_3 (1N) (ml)
30	50	0	35	1.5	10
45	50	0	35	2.5	10
100	25	50	35	4	4
150	25	50	35	12.5	4
200	50	50	35	10	4
250	50	50	35	12.5	4

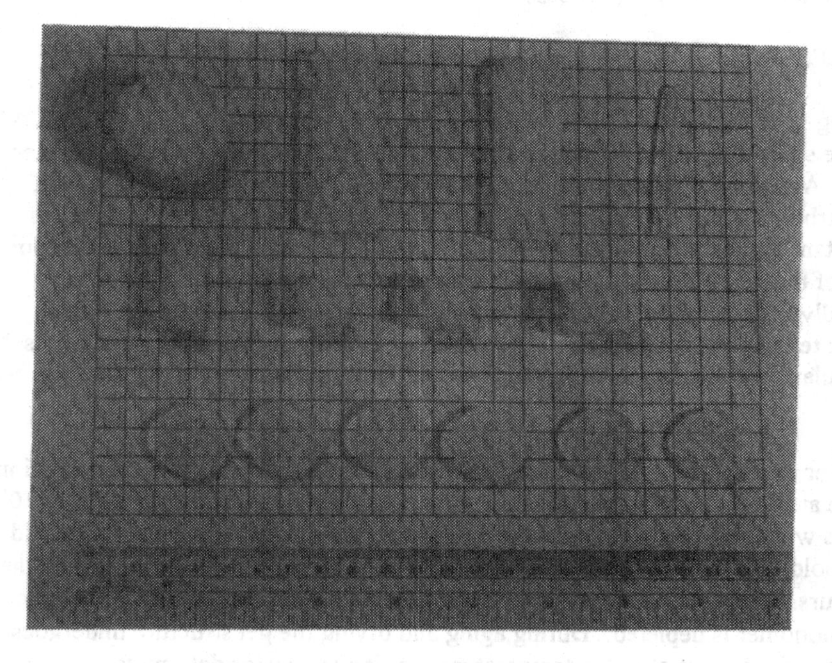

Figure 1. A variety of 20nm pore radius monoliths produced by a one step HF catalyzed process.

For high concentrations of HF, gelation at room temperature can be very rapid, often on the order of a few minutes. Too rapid a gelation of the sol is undesirable for several reasons including insufficient processing time, poor homogeneity and bubble formation. Hydrolysis is exothermic thus the temperature rises rapidly after the addition of TMOS. For the formulation used here the temperature peaks within 2 minutes. In order to increase the time available for filtering and casting, the reactants are generally cooled in an ice bath. The heat generated by hydrolysis is just enough to raise the temperature to approximately room temperature (25 °C) allowing processing and gelation to occur at a stable temperature.

The addition of methanol as a cosolvent has several effects on the sol. It enhances the miscibility of the TMOS and water, raises the pH of the solution slightly, moderates the temperature rise by increasing the total thermal mass of the solution, and slows down the condensation reaction allowing more time for processing the sol. It also increases the total volume of gelation which surprisingly doesn't have much affect on the final gel texture. One major effect of added methanol is that it decreases the solubility of silica in solution. This is an important consideration in the aging process.

PROCESSING

Casting

The sol is allowed to stir for several minutes before being cast into a polymer mold. A variety of plastics can be used as molds. Polystyrene, polypropylene, polycarbonate, and Teflon® have all been successfully used. Hydrophobicity is important in minimizing the formation of a meniscus which often causes the top edge of the gel to break off around its circumference during drying. Gels are normally transferred to Teflon containers for drying and are brought to a final drying temperature of 180 °C. Teflon containers are sometimes used for molds, particularly for large monoliths which are difficult to transfer without damage.

Aging

After casting, the gels are placed on a stable surface in an undisturbed location to age at room temperature for a minimum of 2 days. Significant syneresis (≈10%) occurs within an hour and continues at a lower rate throughout the aging period. The molds are then placed in an oven and aged at higher temperatures for another 24 hours. This aging step strengthens the gel and ensures that all unreacted silicic acid monomer is depleted. During aging and drying the gel structure undergoes Ostwald ripening at elevated temperatures. Aging is a major determinant of the final gel structure (along with the formulation and drying method).

Drying

The drying process used for large pore monoliths is essentially the same as with other gels except the larger pores provide lower capillary stresses and greater permeability allowing the process to be shortened to as little as 24 hours for small monoliths (25mm x 7mm). After aging, the wet gels are carefully removed from the molds, washed with deionized water, and placed in drying containers. Larger monoliths are more difficult to handle and consequently are most often cast and dried in the same Teflon container. The drying schedule varies depending on the size of the monolith. Larger monoliths have a smaller external surface area relative to their volume and a longer diffusion path for the escaping solvent. Thus they require longer drying times.

The maximum stress on the gel occurs at the critical point i.e. the point at which the liquid begins to penetrate the pores. This is also the point at which the gels begin to turn opaque. Once the gel has survived the opaque stage, it is insensitive to the rate of further drying. After this stage is over, the gels are brought a final temperature of 180°C as this is the temperature at which most physically adsorbed water is removed.

Rehydration

The best way to rehydrate a gel is by immersing only the bottom of the gel in liquid. This allows the air in the pore network to escape. Experience with this class of gels has proven that the threshold pore radius for successful rehydration of dried (fully hydroxylated surfaces) gels is in the vicinity of 70 -80Å.[8]

ANALYSIS

The large pore gels in this study were analyzed by nitrogen adsorption, mercury pycnometry and helium pycnometry to determine pore characteristics, bulk density and structural density. Measurements were made on the as dried gels and at several stabilization temperatures up to full densification at 1150°C. High resolution Scanning Electron Microscopy was used to determine the morphology of the dried gels and UV/Visible/NIR spectroscopy was used to determine optical properties using a Perkin-Elmer Lambda 9 spectrometer.

For nitrogen adsorption and pycnometry, the gels were outgassed at 180°C for 12 hours before analysis. Nitrogen adsorption was conducted on the Autosorb-6, manufactured by Quantachrome Corporation, Boyington Beach, Florida. Structural density was accomplished using the QPY-1 helium pycnometer also manufactured by Quantachrome. Due to the ability of large pore monoliths to be easily rehydrated, pore volumes were verified using water pycnometry. Bulk density was determined by mercury pycnometry and cross-checked with BET data. SEM was accomplished on the JSM 6400 Scanning Electron Microscope manufactured by JEOL Inc., Boston, MA.

RESULTS AND DISCUSSION

Pore Texture

Figure 2 shows the nitrogen adsorption data for a representative sample of a 200Å large pore monolith. The BET surface area is 159 m^2/g, and the pore volume is 1.68 cm^3/g giving an average pore radius of 212Å. Figure 3 is a high resolution SEM of a fracture surface showing the colloidal texture of the gel. The BET analysis is virtually identical at all stabilization temperatures up to 1000°C at which point the gel begins to consolidate.

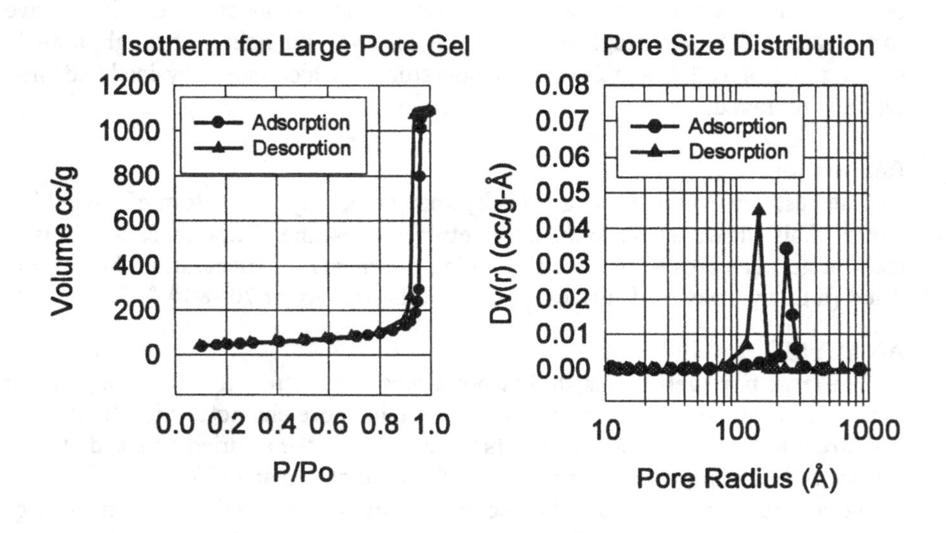

Figure 2. Nitrogen adsorption data for dried large pore monolith.

The stabilization of large pore gels is very similar to their smaller pore cousins except that pore shrinkage and closure occurs at a higher temperature. Due to their lower specific surface areas (typically 160 m^2/g vs 400 m^2/g for a 45Å gel) large pore gel monoliths are less hydrophilic and retain less water than higher surface area gels. A typical 200Å pore radius gel will pick up approximately 2% by weight in water vapor from the atmosphere as opposed to more than 6% for a 45Å gel. The greater permeability of large pore gels also allows water vapor to more readily escape during consolidation. Consequently, the large pore monoliths can be brought to full density under ambient atmosphere without dehydroxylation treatments.

Figure 3. High resolution SEM of dried large pore monolith fracture surface.

Densification.

Full density is normally achieved by heating the gels to 1150°C - 1200°C. Figure 4 shows the pore radius and bulk densities of the large pore gels as a function of stabilization temperature. They are compared to the stabilization properties of similar acid catalyzed gels of smaller pore radius as evaluated by Wallace.[9]

Dried monoliths are machinable. They are relatively soft and can be cut dry with ordinary tools (such as a small band saw, or lathe) or shaped with a grinding wheel. They can then be densified as described above. Thus larger more intricately shaped pieces can be cut from a standard shaped cylinder.

Optical Characteristics

Porous monoliths are transparent but scatter light in the visible and UV according to the size of the pores. Scattering becomes visibly noticeable at a pore radii above 20 nm and progressively increases with pore size. Although pore radii

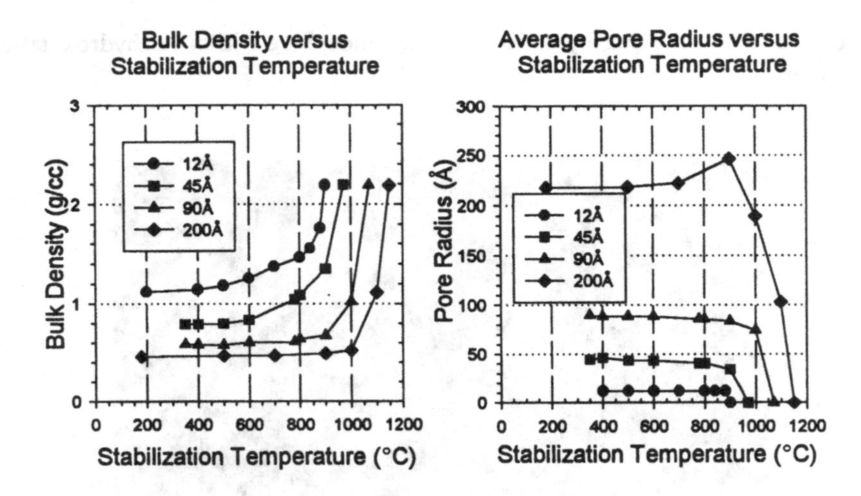

Figure 4. (a) Bulk density versus stabilization temperature for 200Å radius gels as compared to similar gels of smaller radius. (b) Average pore radius of gels versus stabilization temperature. (Small pore data from Wallace[9]).

as large as 40nm have been produced, 20nm radius monoliths retain sufficient transparency for optical applications and possess the most consistent processing parameters. Figure 5 shows the UV-visible spectrum for a 6mm thick dried monolith with an average pore radius of 20nm contrasted with the same sample filled with deionized water. In accordance with Rayleigh theory, scattering decreases dramatically when the pores are filled with solvents of higher refractive index. As with the porosity, there is little change in the optical transmittance of the porous monoliths until sintering begins at 1000°C. Surprisingly, the scattering increases as sintering occurs despite the decreasing pore size. Thus the gels enter another opaque stage during densification. Upon full densification, the gels regain the transparency of dense silica glass as shown in the UV/Vis/NIR spectrum of Figure 5. In this figure the spectrum is plotted with that of an optical flat made of commercial Gelsil®, a high purity (low OH) gel-silica optical glass produced by Geltech Inc. of Orlando, Florida. Without chemical dehydroxylation the dense large pore derived glass still retains significant water which degrades the transparency in the ultraviolet and infrared regions.[10]

CONCLUSIONS

The large pore gel system described here shows promise in several areas such as optical sensor substrates, catalyst supports and as preforms for dense high purity

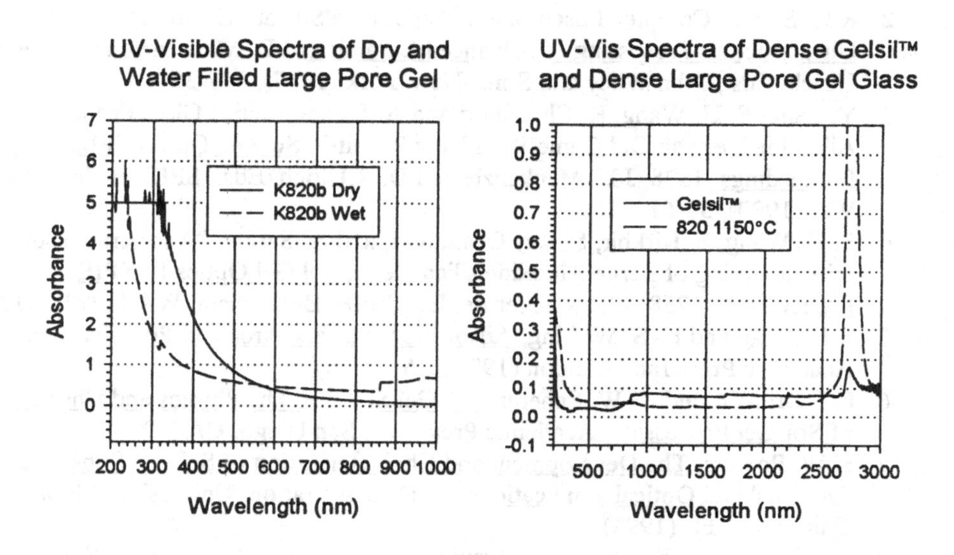

Figure 5. (a) UV/Vis spectra of porous monolith dry as compared to the same sample filled with deionized water. (b) Fully dense sample compared with commercial high purity optical Gelsil®.

silica components. In the smaller size range (up to about 2.5cm by 1cm) they are easily manufactured with excellent yields (approx. 90%). Larger sizes are progressively more difficult to produce. At a pore radius of 200Å, they retain enough transparency to be useful in optical sensing applications, can be easily rehydrated and redried without failure and can be reliably brought to full density. The relatively large pore size imparts good permeability and the large pore volume permits doping with a wide variety of materials.

ACKNOWLEDGMENTS

The financial support of AFOSR under ASSERT Grant No. F49620-94-1-0295 is gratefully acknowledged.

REFERENCES

1. L.L. Hench, G.P. LaTorre, S. Donovan, J. Marotta, E. Valliere, "Properties of gel-silica optical matrices with 4.5nm and 9.0 nm Pores", Sol Gel Optics II, SPIE Proceedings, 1758, J.D. Mackenzie, Ed., , SPIE, Bellingham, WA (1992), 94.
2. R.D. Shoup, Complex Fused Silica Shapes by a Silicate Gelation Process, in Ultrastructure Processing of Advanced Ceramics, J.D. Mackenzie and D. R. Ulrich, Eds., John Wiley and Sons, New York (1987), 347-354.
3. Y. Sano, S. H. Wang, R. Chaudhuri and A. Sarkar, "Silica Glass From Alkoxide Gels; an Old Game With New Results", Sol Gel Optics SPIE Proceedings, 1328, J.D. Mackenzie and D. R.Ulrich, Eds., SPIE, Bellingham, WA (1990), 52-61.
4. S. H. Wang, F. Kirkbir, S. Ray Chaudhuri, and A. Sarkar, "Accelerated sub-critical Drying of Large Alkoxide silica Gels", Sol Gel Optics II, SPIE Proceedings, 1758, J.D. Mackenzie, Ed., SPIE, Bellingham, WA (1992), 113.
5. S.J. Gregg and K. S. W. Sing, Adsorption, Surface Area and Porosity, 2nd Ed, Academic Press Inc., London (1982), 25.
6. C.J. Brinker, and. G.W. Scherer, Sol-Gel Science, The Physics and Chemistry of Sol-Gel Processing, Academic Press Inc., San Diego, CA (1990).
7. K.W. Powers, The Development and Characterization of Sol Gel Substrates for Chemical and Optical Applications, Ph.D. Dissertation, University of Florida, Gainesville, FL (1998).
8. J.K. West, J.M. Kunetz, F. G. Araujo, and L.L.Hench, "Stability of Sol Gel Silica to Water Diffusion: Experimental and Theoretical Analysis", Sol Gel Optics III, SPIE Proceedings, 2288, J.D. Mackenzie, Ed., SPIE, Bellingham, WA (1994), 113-125.
9. S. Wallace, Porous silica Gel Monoliths: Structural Evolution and Interactions With Water, Ph.D. Dissertation, University of Florida, Gainesville, FL (1991).
10. L.L. Hench, S. H. Wang, and J.L. Nogues, "Gel-silica Optics", Multifunctional Materials, Proceedings of the SPIE, 878, R.L. Gunshor, Ed., SPIE, Bellingham, WA (1988), 76-85.

STRUCTURE OF SONO-AEROGELS PREPARED BY A NON-AQUEOUS METHOD

L. Esquivias*, A. González-Pecci, J. Rodríguez-Ortega, C. Barrera-Solano and N. de la Rosa-Fox
Dpt. Física de la Materia Condensada.
Apdo.40, Puerto Real, 11510. Cádiz. Spain

ABSTRACT

A simulation of the structure of dense gels based on the technique of random close packing (RCP) is presented. These gels are formed by monoparticular packing of spheres, according to the TEM observations. The strategy consists of quantifying their structure by the calculation of their pore volume distribution and by the mesopores. These distributions are compared to experimentally calculated distributions. The results show the existence of the hierarchical levels. The models allow calculating the local density associated with each hierarchical level, and packing of the successive levels. This technique has been applied to silica dense sono-aerogels elaborated from tetraethylorthosilicate by non-aqueous reaction.

INTRODUCTION

It is really fortunate that a system of many particles governed by quantum statistical mechanics as the atomic structure of crystal could be represented by a naïve representation of packed spheres. Models of random network spheres also can represent the atomic structure of glasses, amorphous metals and alloys [1]. It seems reasonable to extend this representation to gels since their elementary particles are, at least, one order of magnitude larger and the involved interactions

* Contacting author. E-mail: luis.esquivias@uca.es. Fax #: +34 956 834924.

To the extent authorized under the laws of the United States of America, all copyright interests in this publication are the property of The American Ceramic Society. Any duplication, reproduction, or republication of this publication or any part thereof, without the express written consent of The American Ceramic Society or fee paid to the Copyright Clearance Center, is prohibited.

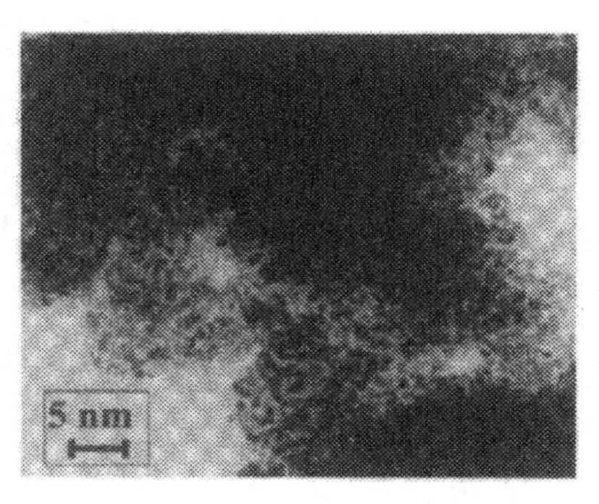

Figure 1. Micrograph of a silica sono-aerogel prepared from TEOS+H_2O suggesting a particulate structure.

are not quantum type. Besides this, modeling of gels has also the advantage that is not necessary to consider any chemical order.

EM observations of sonogels [2] recommend describing them (Figure 1) as a collection of packed spherical particles. This particular structural characteristic, which is induced by the solventless processing and the cavitation phenomenon, facilitate the study of their structure if an adequate strategy is applied. The immediate suggestion is to implement a macroscopic 3-D image of these gels by means of a collection of packed spherical particles. The same is true when Drying Control Chemical Additives (DCCA) or a combination of both external agents is used [3, 4]. The common circumstance is that in all cases we are dealing with dense gels. The texture of dense gels has certain features that permit an approach from simple adsorption data. In these materials, the main contribution to specific surface arises from mesoporosity. This is equivalent to generate a texture local model at a mesoscale level. In a previous work we applied this technique to silica xerogels prepared with DCCA [5] that allowed quantifying their structure from the mesopore fraction on the basis of the interstitial distribution function [6].

We have also applied this technique to Hg porosimetry data of sono-aerogels [7]. These samples are mesoporous and fit properly the RCP models, presenting a hierarchic structure. According to the fitting with our models, overlapping between macro- and mesopores exists in one of those samples. We shown that the viewpoint of this method can be extended to other pore range defining the pore type according to the particle arrangement instead of their size. Thus, we defined type I pores are those interstices between four contacting particles. Type II pores are holes between not contacting particles which are not large enough to contain at least one particle, otherwise, they are type III. This only signifies a scale change that is marked by the resolution of the probe. Certainly, when the probe is Hg intrusion the texture of gel may change but the main goal of this paper is to show that the method we used is valid to represent the structure of dense gels.

However, this does not imply that we give up to profit from the information that can be drawn for the knowledge of the texture in the absence of isostatic pressure.

The functional features and the technological applications of ceramic materials depends on complex relations between their chemical composition, microstructure and processing. The use of a non-aqueous route to gel [8] has been proposed to facilitate the ceramic processing. The combination of a non-aqueous route to gel with the use of ultrasound is unexplored so far. In this paper we present the results on the structure of aerogels prepared from the reaction of tetraethoxysilane and a formic acid under ultrasound. The high volume ratio formic acid/TEOS needed for reaction involves that no dense gels are obtained because the remaining liquid phase after gelation represent an important amount of the total volume. Consequently, the aerogels under study are light. However we will show that, when the probe is Hg intrusion, the isostatic pressure eliminates macroporosity and some information can be drawn.

STRUCTURAL MODELS.

The gel structure is depicted as a collection of packed spherical particles [9]. We have created a catalog of pore size distributions and compared them to that of a series of gels prepared in different conditions to choose the more appropriate model. Their generation is based on the classic models of packing spheres of Scott [10], and Bernal and Mason [11], Finney and Wallace [12] and Frost [13]. In these studies, the topological disorder is described by a histogram of forms and the volume distribution in the porous space. A characteristic behavior is the frequency of distinct spatial forms such as tetrahedra or octahedra. The model characteristics are expressed as a function of particular macroscopic parameters. The most relevant is the packing fraction, C. The pair correlation function P(r) represents the probability of finding a sphere at an r distance from other arbitrary sphere. Finally, the average contact number is calculated as $\eta = \rho \int_0^{d+\varepsilon} P(r) 4\pi r^2 dr$, where d is the diameter (d >>ε) and ρ is the density.

On the models of Scott [10], Bernal y Mason [11] and Finney and Wallace [12] show that η increases almost linearly with C, according to $\eta = 12.7C$. Anyway, C depends on the contact criterion that is defined as function of ratio L (center to center distance)/d.

The coordination number depends on the compactness. When voids are randomly introduced into the structure C decreases as expected. Curiously, it does not affect P(r), simply causing flattening of the peaks but the main features stand. This is not the case of η, which drops from 9 to 6 when 35% of extra porosity is added. Consequently, although it would be possible to deduce P(r) from an experiment, this is not useful, as it would be imperceptible to the structure. C turns out to be a better reference, although it is influenced by the contact distance criterion.

To check these models is necessary to establish a link between the "particle space" and the "pore space" from where the most of the information on the structure of gels is drawn. Finney and Wallace [12] and Frost [13] determined the distribution of spheres of largest radius inscribed in the interstices of the models

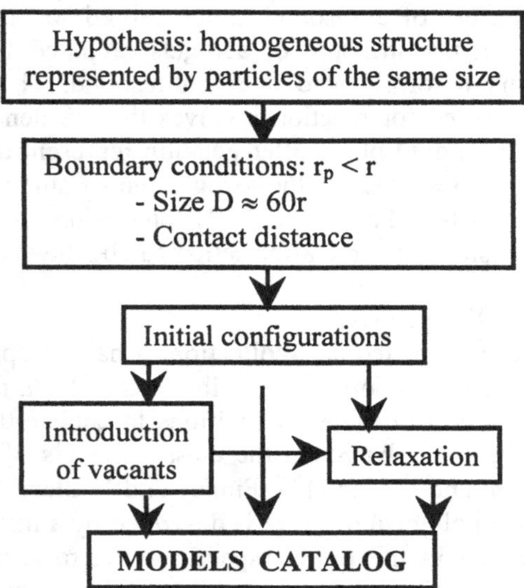

Figure 2. Outline of the process followed to generate our catalog of models.

of Scott [10], and Bernal and Mason [11]. From here, a function that allows quantitative evaluation of the pore size distribution can be obtained experimentally.

DESCRIPTION OF THE METHOD.

The catalog was generated from three basic spherical models, H, D and S of diameters, $D \sim 60r$ according to the outline presented in Figure 2. Table 1 accounts for the closeness of contact conditions and resulting compactness of each case. The model pore volume distributions, p(K) were calculated from the largest sphere radius inscribed within the interstices [13] and expressed as a function of a reduced variable K ($K = r_p/r$, r_p being the pore radius and r the particle radius). In this way, the results could be extrapolated to whatever particle size. At this point it is worth to remind that the p(K) functions provide a quantitative way of evaluating pore-size distribution without implying necessarily spherical pores. The inscribed larger spheres constitute rather a local gauge of the channels which are formed from a succession of interstitial sites and which may be linked by constrained necks [14].

The S model, as in Scott's loose model [10], should be mainly formed by octahedra, slightly distorted, because the maximum of the distribution is near $K = K_{max} = 0.414$, which corresponds to octahedral sites. The D model has the maximum population of pores for $K_{max} = 0.34$. This is a consequence of allowing interpenetration or deformation of the spheres. The distribution is intermediate between the Scott's [10] and Finney's[12] models. The H model allows 10% of radial deformation giving rise to a model with the maximum of the distribution $K_{max} = 0.25$, not far from that corresponding to the tetrahedral sites ($K_T = 0.225$). Finney's model has K_{max} at 0.3, although it is a little more compact, the distribution ranging from 0.225 to 0.8.

The models are assisted to converge toward a particular structure introducing random voids into the structure and relaxing the network applying the Lennard-Jones potential [12]. A Monte Carlo algorithm finds the distribution of minimum energy. The models are identified according the following nomenclature: *X*P*y*L*z*

X indicates the type of model: H, D, S.
y is the percentage of added porosity
$z = 10 \cdot \sigma$ parameter Lennard-Jones potential, i.e., L8 means $\sigma = 0.8$.
LJ indicates a non-relaxed model.

Table 1. Contact criterion and models compactness.

Type of model	Closeness of contact (L/d)	C
H	$0.91 < L/d < 1.00$	0.55
D	$0.95 < L/d < 1.02$	0.52
S	$1.00 < L/d < 1.18$	0.44

The models' volume distribution derivatives $\psi(K) = K^2 p(K)$ were traced as a function of log K. In Figure 3 are represented some of them. This allows, when compared to that of a series of gels (measured by means of N_2 adsorption method or Hg intrusion), to choose the more appropriate model, simply sliding it along the r_p axis until the position considered to give the best fit.

The gel structure could be arranged at several hierarchic levels [9]. In this case, the spheres of the $i+1^{th}$ level, r_{i+1} radius, are formed by a similar arrangement of smaller spheres of r_i radius, q_{i+1} being the radius ratio, $q_{i+1} = r_{i+1}/r_i$. These spheres could be also formed by random close packed spheres of $r_{i-1} = r_i/q_i$ radius (Figure 4). The upper value of *i* is limited by the gel's apparent density. K at the *i-th* level, K^i, is related to the pore size, r^i_p and elementary particle size, r_0, by the relationship

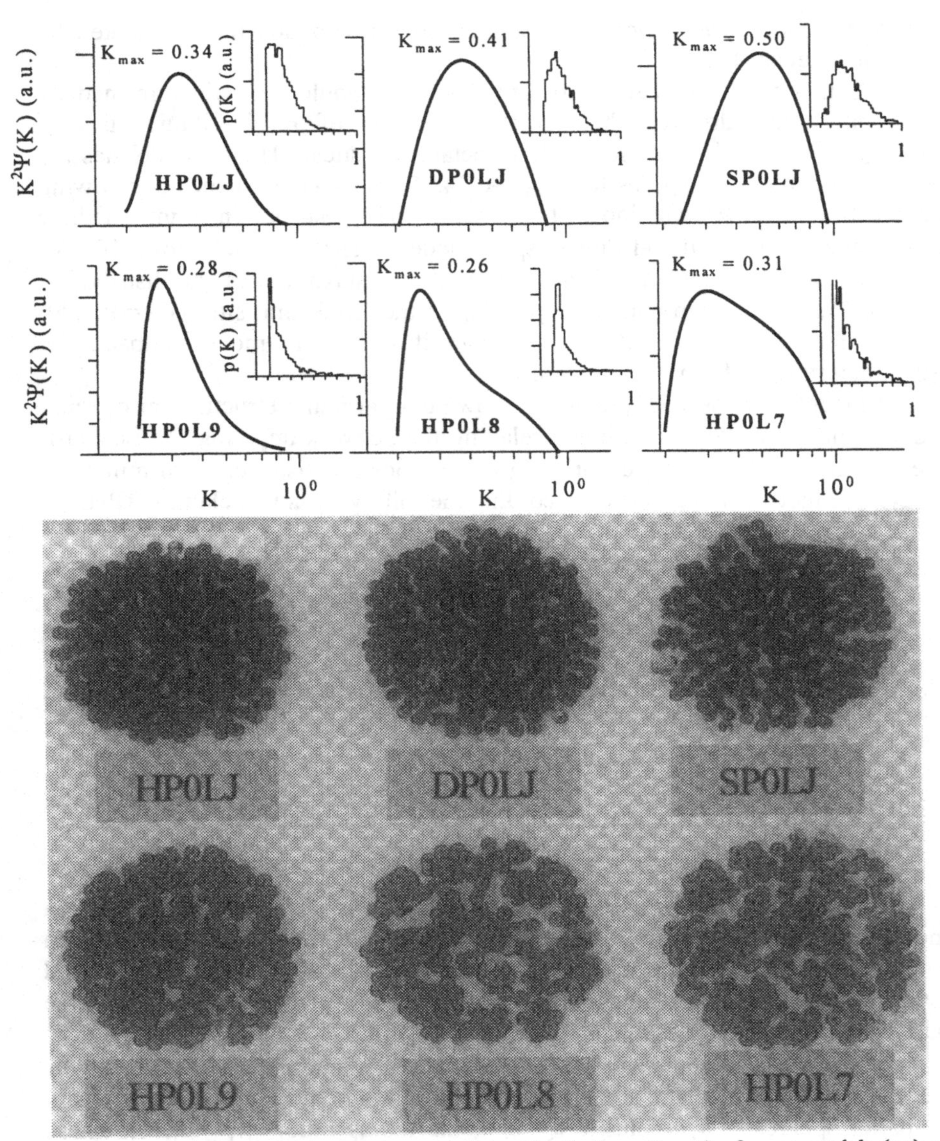

Figure 3. Pore volume derivatives and pore volume distributions (insets) of some models (up) and their bi-dimensional representations (down).

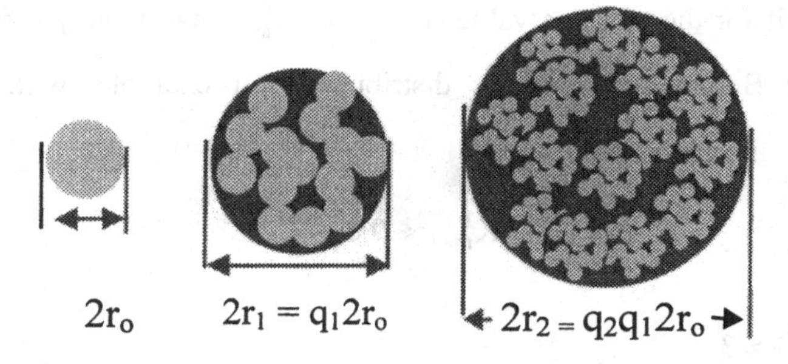

Figure 4. Representation of the hierarchic structure.

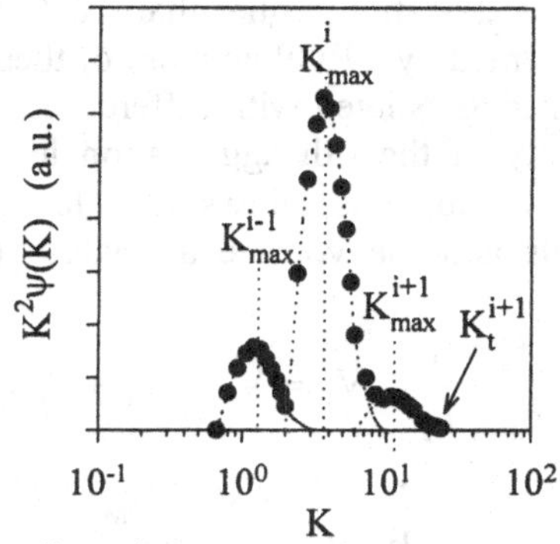

Figure 5. Pore distribution outline of a structure formed by successive packing of spherical particles built of smaller particles random close packed.

$$K^i = \frac{r_p^i}{r_i} = \frac{r_p^i}{\prod_1^i q_i r_o} \qquad (1)$$

We use an estimated value of K_{max} as a criterion to choose the more appropriate model from our catalog (Figure 5). We assume that $\frac{r_{pmax}^0}{r_{pt}^0} = \frac{K_{max}^0}{K_t^0} = a$. r_{pt}^i is the position of the distribution tail.

Therefore, from equation (1),

$$K_{max}^i = \frac{r_{pmax}^1}{\prod_1^i q_i a r_{pt}^o} K_{max}^o \qquad (2)$$

We admit for the analysis values of a and K^0_{max} that of very close networks such as the Bernal and Finney's distribution, for example, with $a = \frac{3}{8}$ and $K^0_{max} = 0.3$. In such a case

$$K^1_{max} \approx 0.8 \frac{r^1_{pmax}}{r^0_{pt}} \tag{3}$$

with $q_1 \approx \frac{r^1_{pt}}{r^0_{pt}} \approx 3$

We take this value just as a standard, bearing in mind other plausible hypotheses would not give a deviation higher than 10%. In the next analysis we consider the gel structure formed by a RCP network of these clusters.

Data on the pore volumes associated with different hierarchical levels, size of aggregates, the local density of the i-th aggregation level, and other structural parameters [5]. This set of equations allows to calculate the densities at the different aggregation levels and the volume at macro- (V_M), meso- (V_m) and microscale (V_μ)

$$\frac{1}{\rho_a} - \frac{1}{\rho_s} = V_T = V_M + V_m + V_\mu \tag{4}$$

$$\frac{1}{\rho_2} - \frac{1}{\rho_s} = V_T - V_M \tag{5}$$

$$\frac{1}{\rho_1} - \frac{1}{\rho_s} = V_T - V_M - V_m \tag{6}$$

From Hg porosimetry the apparent bulk density, ρ_a, and corrected bulk density, ρ_{corr}, are obtained. The former refers to the external volume of the solid itself, $1/\rho_a = V_T + 1/\rho_s$. $1/\rho_s$ being the specific volume occupied by the solid phase. The second one gives the volume filled by the Hg, $V_{Hg} = 1/\rho_{corr} - 1/\rho_s$.

EXPERIMENTAL

We have tested our models with some SiO_2 gels elaborated by reaction of TEOS with formic acid using ultrasounds with 6 [TEOS]/[HCOOH]. The reaction took place in an open container dissipating an ultrasound energy density 10 $J \cdot cm^{-3}$. The device delivered to the system 0.6 $w \cdot cm^{-3}$ of ultrasound power [2]. The total dissipated was 150 $J \cdot cm^{-3}$. The homogeneous liquids were poured in glass hermetic containers at 50°C until gelation. Hypercritical drying [15] of the wet gels in an autoclave (p=120 bar and T=250°C) leads to monolithic pieces of aerogels.

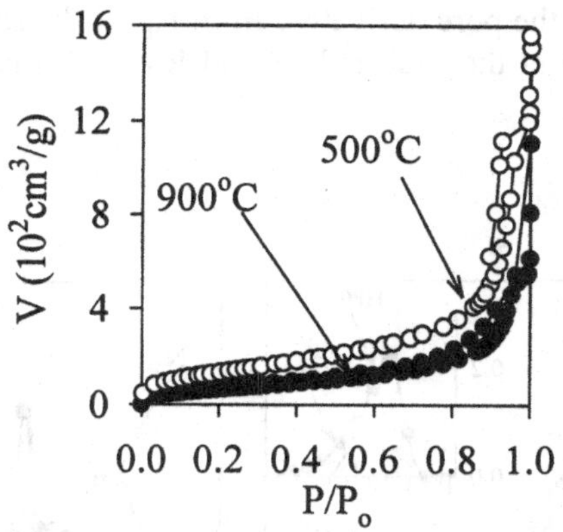

Figure 6. Isotherms of the sample heat treated successively to 500°C and 900°C. They are type II which is characteristic of macroporous solids presenting a hystersis loop type H 3.

They were successively heat-treated at 500°C for 10 h in Cl_4C atmosphere and at 900°C in air for 3, that we call STEF6-5 and STEF6-9, respectively. They were texturally characterized by isothermal nitrogen adsorption in an automatic device. Horvath-Kawazoe method [16] has been employed to calculate the pore size distributions from adsorption branch. Porosity was also characterized by mercury intrusion running until a maximum pressure of 340 MPa.

APPLICATION OF THE MODELS

The isotherms are Type II from IUPAC [17] classification (Figure 6). They respond to solids with a wide pore distribution, including the macropores range. It makes not feasible the application of our mesoporous models. The hysteresis loop that shows the STEF6-5 sample is Type H2, which features particulate materials, presenting narrow necks and wide cavities. The sample heat-treated at 900°C exhibits an aspect Type H3 that features absorbent materials constituted by plate-like particles. Nevertheless, the pore volume distributions from N_2 of both samples are similar. Hg intrusion reduces macroporosity and densifies the sample, displaying pore volume distributions which are quite different and to which is possible to apply our models (Figure 7). TEM micrographs do not feature particulate dense gels (Figure 8). The sample after heat-treatment at 900°C became stiff enough to allow the mercury to reach a lower level of particles. However, the heat-treatment also causes particle aggregates deformation and occlusion of the material external surface.

Suitable fitting was accomplished by applying the models HP0LJ (K_{max}=0.30) and HP0L9 (K_{max}=0.28), to the pore volume distributions from Hg intrusion of the sample heat-treated at 500°C, respectively. The models HP0L9 (K_{max}=0.28) and

DP0L10 (K_{max}=0.38) fit the pore derivative from Hg of the sample heat-treated at 900°C. The particle size at the first and second levels (r_l) is estimated from the models K_{max}.

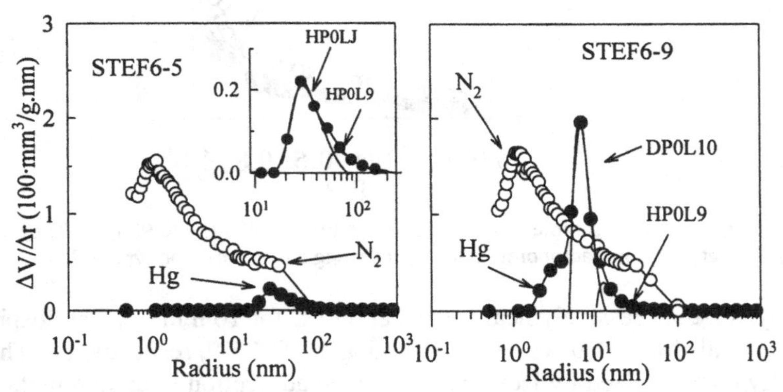

Figure 7. Pore size derivatives obtained from both N2 physisorption (O) and Hg intrusion porosimetry (●). Mesopore fractions of both gels from Hg porosimetry fitted with the models indicated. The inset is a magnification of the pore size derivative from Hg.

We suppose for the backbone the density of the vitreous bulk silica, ρ_s = 2.2 g·cm^{-3}. From the mesoporous volume measured by physisorption, V_m, we obtain ρ_2, since the isotherms indicate the absence of micropores, $\rho_1 \approx \rho_s$. The corrected densities are below ρ_2. This means that the mercury cannot reach the second level (which is the first from the Hg intrusion point of view). ρ_2 for the STEF6-5 sample is near the double of that of the STEF6-9 sample since the fine structure of the gel calcined at 500°C (Figure 8) collapses. This gives rise to agglomerates of R_l = 250 nm average radius constituted by elementary particles of 100 nm radius (Figure 9). However, the heat-treatment at 900°C has made the sample stiff enough to allow the elimination of the macropores and increasing mesoporosity without collapsing the structure. The Hg penetrates into the structure up to the mesopore level. At this level the particle radius under pressure is r_l = 16 nm forming aggregates of 46 nm radius (Figure 9). With the help of Hg porosimetry we are able to measure the porosity defined by the particles, not by the "particles + structure" because the pore volume derivative fit our models. r_l is the particle size at the mesoscale level that we are not able to detect by N_2 physisorption because it is masked by the structural porosity.

Figure 8. TEM micrographs of both samples under study. STEF6-5 (right) and STEF6-9 (left).

Table 2. Data obtained from the analysis.

	V_{Hg} (cm^3/g)	V_{N2}[†] (cm^3/g)	ρ_a (g/cm^3)	ρ_{corr} (g/cm^3)	ρ_2 (g/cm^3)
STEF6-5	4.3	1.92	0.12	0.25	0.42
STEF6-9	1.5	0.88	0.16	0.21	0.75

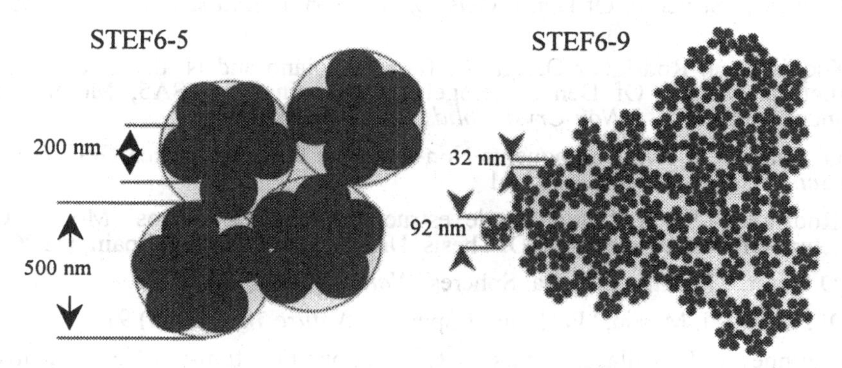

Figure 9. Representation of the STEF6-5 (left) and STEF6-9 (right) gels textures.

CONCLUSIONS

STEF6 samples present a wide pore size distribution, including the macropores range that makes not feasible the application of our models

For the STEF6 samples, Hg porosimetry is not appropriate to determine the pore volume at different levels.

However, in samples stiff enough it is possible to propose models for the STEF6 samples structure from Hg porosimetry data since the Hg reaches the mesopore level.

[†] Measured by N_2 physisorption.

For the sample heat-treated at 900°C we measure the particle size at the mesoscale.

REFERENCES

1 J. Blétry, "Spheres And Distances Models For Binary Disordered Systems" *Phil. Mag. B* **62** (5) (1990) 469-508.

2 N. de la Rosa-Fox, L. Esquivias and J. Zarzycki, *Diffusion and Defect Data*, "Glasses from Sonogels" **53-54**, (1987) 363.

3 N. de la Rosa-Fox, L. Esquivias and J. Zarzycki, "Silica Sonogels With Drying Control Chemical Additives" *J. Material. Sci. Let.* **10**, (1991) 1237.

4 E. Blanco, M. Ramírez-del-Solar, L Esquivias, AF Craievich. "Processing Of Silica Xerogels Using Sonocatalysis And An Additive" *J. Non-Cryst. Solids*, **147&148**, (1992) 296.

5 J. Rodríguez-Ortega and L. Esquivias, "Structural Models Of Dense Gels" *J. Sol-Gel Sci. &Tech.* **8**, (1997) 117.

6 J. Zarzycki, "Structure Of Dense Gels" *J. Non-Cryst. Solids*, **147&148**, (1992) 176.

7 L. Esquivias, J. Rodríguez-Ortega, C. Barrera-Solano and N. de la Rosa-Fox, "Structural Models Of Dense Aerogels" Proceedings of ISA5, Montpellier (France) Sept 1997. *J. Non-Cryst. Solids*, **225**, (1998) 239-243

8 K. G. Sharp, "A Two Components Non-Aqueous Route To Silica Gel" *J. Sol-Gel Sci. & Technol.* **8** (1994) 35-41

9 J. Rodríguez-Ortega, "Modelos de estructura de geles densos (Models Of Structure Of Dense Gels)" Ph. D. Thesis, Universidad de Cádiz, Spain, (1996)

10 G.D. Scott, " Packing of Equal Spheres" *Nature*, **188**, (1960) 908.

11 J.D. Bernal y J. Mason, "Packing of Spheres" *Nature* **188**, (1960) 910.

12 J.L. Finney y J. Wallace, "Interstice Correlations Functions; A New Sensitive Characterisation of Non-Crystalline Packed Structures" *J. Non-Cryst. Solids*, 43, (1981) 165.

13 H. Frost, ONR Technical Report No. 6 (Division of Applied Sciences, Harvard Univ., Cambridge, MA, 1978).

14 See Fig. 8 in ref 8.

15 M. Prassas, J. Phalippou and J. Zarzycki, "Synthesis of Glasses from Gels: The Problem of Monolithic Gels" *J. Mat Sci.* **19** (1984) 1656.

16 G. Horvath and K. Kawazoe, "Method For The Calculation Of Effective Pore Size Distribution In Molecular Sieve Carbon" *J. Chem. Eng. of Japan,* **16** (6), (1983) 470.

17 K.S.W. Sing, D.H. Everett, R.A.W. Haul, L. Moscou, R.A. Perotti, J. Rouquerol and T. Siemientewska "Reporting Physisorption Data for Gas/Solid Systems" *Pure & Appl. Chem* **57** (4) (1985) 603-619.

Basic and Applied Sol-Gel Science and Processing of Ceramics and Composites

THE PENTACOORDINATE SPECIES IN FLUORIDE CATALYSIS OF SILICA GELS

Kevin W. Powers
Department of Materials Science and Engineering
University of Florida, Gainesville, FL 32611-6400

Larry L. Hench
Department of Materials
Imperial College of Science, Technology and Medicine
Prince Consort Road, London SW7 2BP

ABSTRACT

An ion specific fluoride electrode is used to monitor free fluoride concentrations in HF catalyzed sols as silicic acid is added in the form of tetramethoxysilane (TMOS). It is found that fluoride is rapidly bound by the silicic acid in a ratio of approximately four to one, indicating the formation of multifluorinated silicon complexes. The decrease in pH as TMOS is added provides evidence that anionic hypercoordinate species are formed that are more stable than previously thought. A polymerization scheme is proposed that explains the hydrophobicity of fluoride catalyzed gels and the difficulty in retaining structural fluoride in HF catalyzed sol gel glasses.

INTRODUCTION

The introduction of fluoride into sol gel glasses has several important applications. Fluoride can be used to replace surface silanol groups as a method of dehydroxylation. This enables fluoride treated sol gel glasses to be sintered into high quality (low water) silica glass without bloating and eliminates the tendency for reboil that occurs with some chlorine-based dehydroxylation treatments. Fluoride also is an important dopant for reducing the refractive index of silica glasses such as in the production of cladding for optical fibers. Finally, fluoride has long been known to have a marked catalytic effect on the sol gel reaction under acidic conditions.

To the extent authorized under the laws of the United States of America, all copyright interests in this publication are the property of The American Ceramic Society. Any duplication, reproduction, or republication of this publication or any part thereof, without the express written consent of The American Ceramic Society or fee paid to the Copyright Clearance Center, is prohibited.

It has been theorized that fluoride serves to expand the coordination sphere of the silicic acid monomer (or oligimer) making it more subject to nucleophilic attack. [1] A pentacoordinate fluorinated species has often been proposed as the intermediate [2,3] This experiment presents direct experimental evidence for the existence of a stable anionic hypervalent silicon species involved in the fluoride catalyzed gelation process under acidic conditions.

Pope and Mackenzie [4] demonstrated that with a TEOS precursor, hydrofluoric acid promoted gelation nearly ten times faster than most mineral acids or bases and 100 times faster than the uncatalyzed reaction. The texture of dried HF catalyzed gels is generally characterized by a large pore volume and low surface area, similar to those produced in base catalyzed gels. Thus it is often theorized that the fluoride ion acts similarly to the hydroxide ion in promoting gelation. [5] This argument is further supported by the fact that both species are highly electronegative and of similar size. The hydroxide ion has a radius of 1.40 Å and the fluoride ion is slightly smaller at 1.33Å. Unfortunately, it is difficult to directly compare the effects of fluoride and hydroxide in aqueous solution since the latter is always present at a concentration which is determined by the pH of the solution. Thus the effect of fluoride is best studied under acidic conditions.

Silicon is known to form primarily tetrahedrally coordinated compounds, however it does form a hexacoordinate species with excess HF in aqueous solution, fluosilicic acid (H_2SiF_6). It is the occurrence of this species that undoubtedly prompted the original speculation that siloxane polymerization was due to a hypervalent intermediate. This view was reinforced by Davis and Burggraf who used molecular orbital calculations (MNDO) to demonstrate that pentacoordinate silicon intermediates are energetically feasible in the gaseous state. [6-8] They concluded that small anions can quite easily increase the coordination sphere of silicon through the involvement of the empty d orbitals. This expansion makes the silicon center more susceptible to nucleophilic substitution. Later they applied these calculations specifically to base and fluoride catalyzed polymerization reactions between silicic acid species.

Recently, Hayakawa and Hench [9] performed AM-1 calculations to determine the stability and geometry of fluoride substituted silica rings. These calculations indicate that fluoride can form a stable petacoordinate species with trigonal bipyrimidal geometry on four membered silica rings. Figure 1 shows the optimized geometry of such a structure calculated using AM-1 on an Oxford Molecular Cache workstation. The stability of the structure demonstrates that pentacoordinate silicon centers may be possible on the surface of growing colloidal particles catalyzed by fluoride.

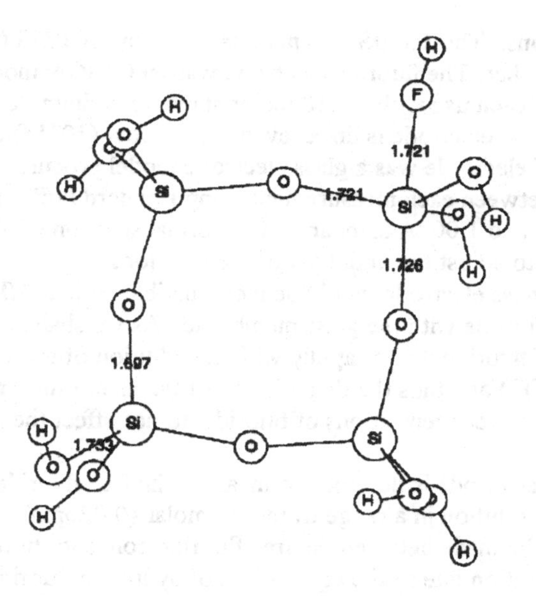

Figure 1. AM-1 optimized geometry shows a pentacoordinate silicon in a trigonal bipyramidal geometry upon attack by one HF molecule. The ΔH_f is -1094.2 kcal/mole, 23 kcal more stable than the same ring with an unassociated HF molecule.

In this experiment an ion-specific fluoride electrode is used to monitor free fluoride concentrations in HF catalyzed sols as silicic acid is added in the form of tetramethoxy silane (TMOS). It is found that fluoride is rapidly bound by the silicic acid in a ratio of four to one indicative of the formation of multifluorinated silicon complexes. The change in pH during the titration provides evidence that at least one anionic hypercoordinate species is formed that is more stable than previously thought. A polymerization scheme is proposed that explains the hydrophobicity of fluoride catalyzed gels and the difficulty in retaining structural fluoride in HF catalyzed sol gel glasses.

MATERIALS AND METHODS

All experiments were conducted in Teflon reaction containers. The hydrofluoric acid solutions were made using Fisher brand 49% HF concentrate diluted in a polypropylene volumetric flask. HF solutions were buffered with TISAB II to a pH of ≈5 for calibration using the fluoride electrode to establish true concentrations. The TMOS was manufactured by ACROS (99%) and procured from Fisher. The fluoride electrode was an ORION model 96-09. It was calibrated before each use with a 0.10 molar standard sodium fluoride solution. The response of the electrode is linear even to very low (10^{-6}M) concentrations of fluoride. The pH electrode was a glass electrode model procured from Fisher. It was calibrated between each measurement using standard buffer solutions of pH 7.00, 4.00, 2.00, and 1.00. A standard 1.00 Normal solution of nitric acid from Fisher was used to adjust the initial pH of the solutions.

Ordinarily a glass electrode would be incompatible with an HF solution due to the reaction of fluoride with the glass membrane. As we shall see, however, the concentration of fluoride drops rapidly with the addition of even a small amount silicic acid (TMOS) and thus the degradation of the electrode is minimal and of short duration. Low concentrations of fluoride do not affect the use of this electrode. [10]

The fluoride electrode is designed to measure the free fluoride ion concentration in solution in a range from 10^{-6} molar (0.02ppm) to saturated. A distinction must be made between the free fluoride concentration and the total fluoride concentration due to the equilibrium of hydrogen fluoride. HF is a weak acid with an acid dissociation constant of $Ka = 6.80 \times 10^{-4}$. The total (initial) fluoride concentration is known.

$$HF_{(aq)} \rightleftharpoons F^{-}_{(aq)} + H^{+}_{(aq)} \quad (1)$$

Fluoride not accounted for by either F^- or HF species must be bound in some manner by the silica in solution. Possibilities include both charged and uncharged species in a variety of hydroxide/fluoride combinations as illustrated in Figure 2.

Removal of the fluoride from solution by silica complexation reduces the amount of free fluoride detected by the electrode. Although the resulting equilibria may be complicated, the fluoride concentration will still be directly related to the pH through the HF/F^- equilibrium. The experiment is designed to determine *the fate* of the fluoride introduced as a catalyst in the sol gel reaction. The fluoride electrode is used to monitor the concentration of free fluoride as silicic acid (in the form of TMOS) is added to an HF/HNO_3/water mixture. The concentrations of HF and HNO_3 were selected to correspond with the formula used for the production of sol gel monoliths with large mesopores.[11]

RESULTS AND DISCUSSION

The primary question is whether the fluoride introduced as HF remains in the form of *HF* (i.e. the equilibrium of Equation 1), free fluoride (F^-), or a silicon complex of some type. Dimerized (H_2F_2, HF_2^-) species can be neglected due to

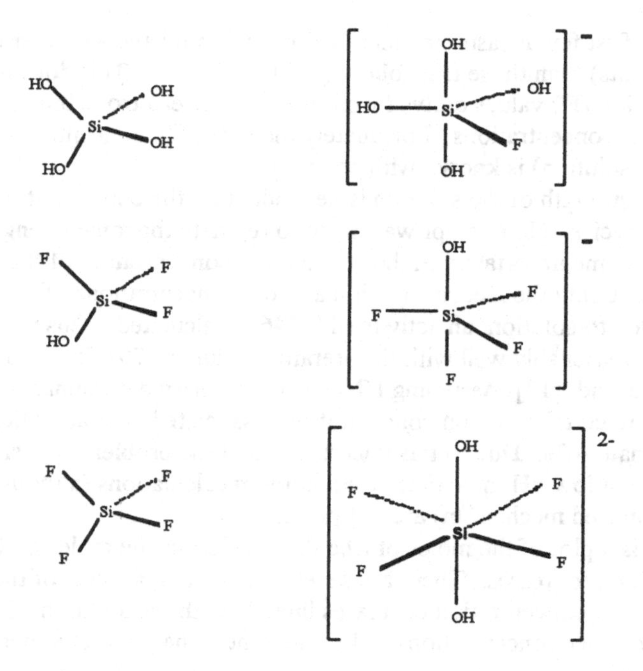

Figure 2. Several of the possible combinations of fluoride and hydroxide around the silicon center with tetrahedral, trigonal bipyramidal and octahedral symmetry.

their very low concentration relative to HF. [10] The fluoride ion concentration can be measured directly with the fluoride electrode. The hydrogen ion concentration is determined by measuring the pH. Any fluoride not accounted for by the HF/F^- equilibrium is assumed to be complexed by the silica present in solution.

The second question is, how does the fluoride complex with silica? Does it form a tetrahedral complex through substitution, or one of the hypervalent complexes discussed earlier? The pH of the solution can be used to give some indication of the type of complex formed.

The temperature of the solution varied by less than 1°C throughout the titration. TMOS was added in 0.1 ml increments and potentiometric readings were taken of the fluoride concentration and pH after each addition. The fluoride reading stabilized in a few seconds at low concentrations of TMOS but took up to several minutes to stabilize toward the endpoint of the titration. The pH readings tended to drift slightly at higher fluoride concentrations but were very stable after the first few data points (at low fluoride concentrations). The pH electrode was recalibrated several times during the procedure to ensure maximum accuracy.

Even so, the first few measurements of pH in each run have a greater uncertainty (±0.05 pH units) than those that follow (± 0.01 pH unit). This does not affect the free fluoride ion (F^-) values derived from the fluoride electrode but may affect the calculated HF concentrations. Fortunately the initial HF concentration (total HF added to the solution) is known with certainty.

The ionic strength of the solution is dependent on the concentration and charge of charged species. No attempt was made to regulate the ionic strength and thus there will be some uncertainty in the activity and concentration of the hydrogen ions present. Using the Debye equation and the concentrations of HNO_3 and fluoride added to solution, an activity of 0.746 is calculated. This value corresponds reasonably well with the literature value of .791 for a 0.1 molal value of pure nitric acid. [12] Assuming HNO_3 is the primary determinant of ionic strength the uncertainty in ion concentrations associated with activities should be no greater than ±6%. Due to this uncertainty and the problems in using the glass pH electrode at low pH, quantitative equilibrium calculations (although consistent with the proposed mechanism) are not presented.

Figure 3 is a plot of the moles of free fluoride ion vs the moles of TMOS added as the titration progresses. There are several interesting aspects of the graph. The free fluoride ion concentration decreases linearly with the addition of TMOS especially at lower concentrations. The endpoint is the point at which the fluoride

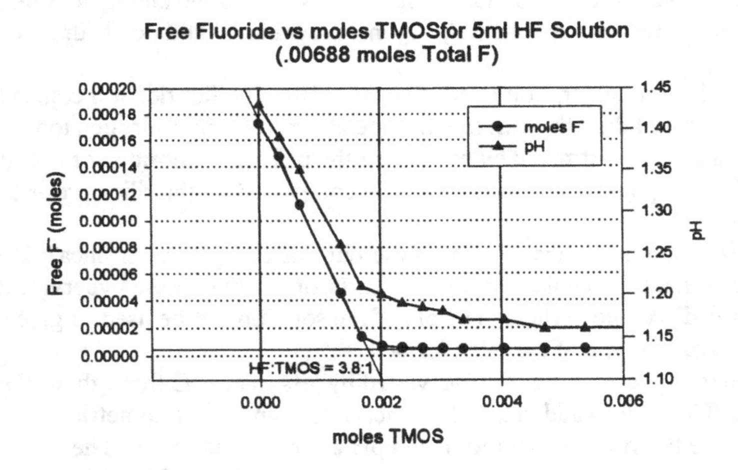

Figure 3 If the decrease in fluoride concentration is extrapolated to zero, the complexation ratio of (total) fluoride:TMOS is determined by the x-intercept.

ion concentration levels off. The fluoride ion concentration does not level off at zero but rather at a low concentration which appears to be due to an equilibrium with the pH and the silicon species that is/are formed. By extrapolating the linear portion of the curves the complexation ratio can be determined. For this concentration of HF (.1075M) fluoride complexes with silicic acid in a ratio of approximately 4:1. The data in Figure 3 clearly demonstrates the ability of silicon to complex fluoride in solution, but this has been long established. What is interesting is the evidence that a hypervalent species is formed that incorporates the hydroxide or oxide ligand. This indicates that there is a competition between the hydroxide and fluoride ligands for the silicon center -- even at low pH. The evidence for a mixed ligand stable hypervalent species is twofold. *The fluoride complexes with the silicic acid in a 4 to 1 ratio after which the fluoride ion concentration stabilizes. Secondly, the pH drops as TMOS is added and the endpoint nears after which it too stabilizes.* The final pH is lower than that which can be accounted for by the added nitric and hydrofluoric acid combinations. The only possible source for the additional hydrogen ions is a hypercoordinate silicic acid complex. If fluoride simply substitutes one for one with the hydroxide in tetrahedral coordination, the hydroxide ion released would immediately combine with a free hydrogen ion to form water and the pH would remain stable. The only way for the pH to drop is for the silicic acid to form a *negatively charged* hypervalent species. Figure 4 depicts this process.

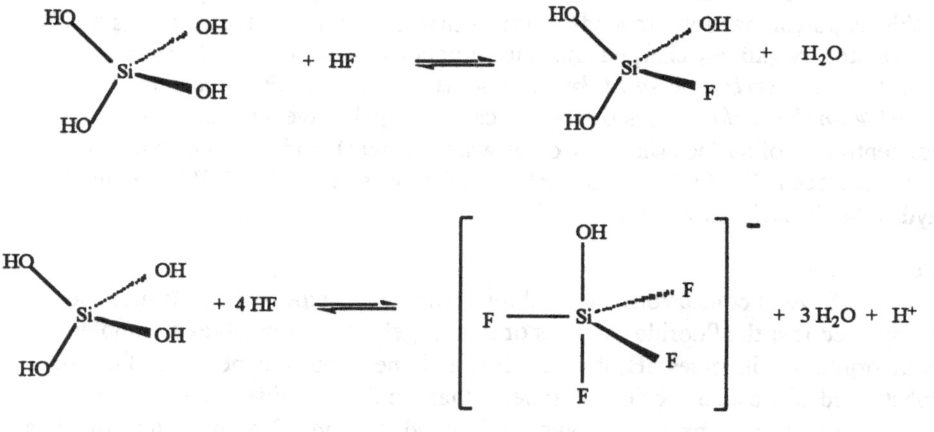

$$Si(OH)_4 + HF \rightleftharpoons Si(OH)_3F + H_2O$$

$$Si(OH)_4 + 4\,HF \rightleftharpoons [SiF_4(OH)]^- + 3\,H_2O + H^+$$

Figure 4. Examples of possible reactions of HF with silicic acid. Substitution reaction (top) produces no change in pH. Reactions that result in charged hypercoordinated species (penta- or hexacoordinated) result in a pH drop.

The case for the participation of the hydroxide ligand in this hypervalent complex is a bit more subtle but still compelling. Figure 3 indicates that the fluoride:silicic acid ratio is in the vicinity of 4:1. Thus nearly all of the silica present in solution has four fluoride ligands. If the species formed is negatively charged and pentacoordinate as is indicated by the pH drop, the fifth ligand must be hydroxide.

It is probable that the multiple substitution of fluoride on the silicon center makes it susceptible to nucleophilic attack by water. In this pH range the concentration of OH^- is less than $1x\ 10^{-12}$ moles/liter. It is small wonder why polymerization occurs under these conditions. As additional silicic acid is added beyond the endpoint, the pH and F^- concentration remain relatively stable. This implicates a single (or at least a dominant) species governing the F^- equilibrium in solution. Since the predominant species present at the endpoint is tetrafluorinated, the hydroxytetrafluoride silicic ion (SiF_4OH^-) is a logical candidate. As siloxane bonds are formed, this species is replaced by a fluorinated surface which (as evidenced by the stable pH) remains pentacoordinate and negatively charged.

If the SiF_4OH^- ion is the operative species, a hexacoordinate surface silicon intermediate is indicated for the condensation reaction. Figure 5 depicts a possible mechanism.

As fluorides are displaced through the formation of siloxane bonds they rapidly combine with the nearest available non-(fluoride)-saturated silicic acid molecule. This results in the fluorides residing at the surface of the growing colloidal particles. The fact that the equilibrium fluoride concentration and pH remain stable as polymerization proceeds suggests that the surface of the colloid is highly fluorinated, negatively charged and is itself pentacoordinated. In excess silicic acid therefore *polymerization will take place so as to maximize the concentration of fluoride on the surface*. This is why HF catalyzed gels have a much lower concentration of surface silanols (lower water content), and generally a much lower surface area. In fact, gels catalyzed with large amounts of HF have very hydrophobic surfaces when dry. [11]

Conclusions

Several conclusions can be drawn from these experiments. It must be recognized that the fluoride catalysis of the sol gel reaction involves a complex equilibrium that involves pH, fluoride, HF, and one or more hypervalent fluoride substituted silicic acid species. It appears that the free fluoride ion is not the operative species in the promotion of rapid condensation. Examination of the data indicates that a hypervalent multifluorinated species is most probably the key to the rapid condensation and that the reaction occurs at the surface of the rapidly

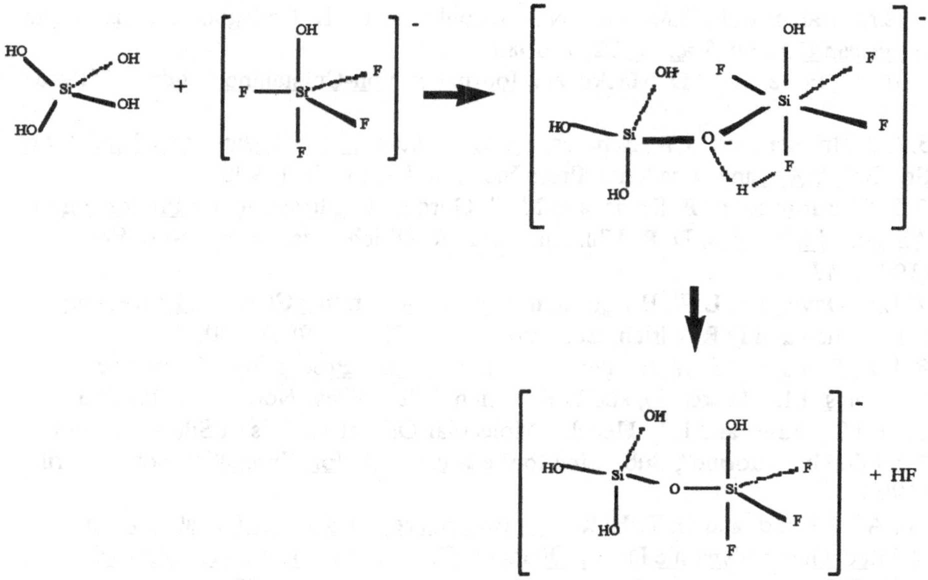

Figure 5. Proposed mechanism for HF catalyzed condensation involving a stable pentacoordinate species with a hexacoordinated intermediate.

growing silica colloids. The pentacoordinate SiF_4OH^- species is a good candidate although more evidence is needed to confirm this. If the mechanism proposed in Figure 5 is correct, the polymerization takes place on the surface of the growing colloid which is highly fluorinated, negatively charged, and pentacoordinated. Molecular orbital calculations such as that depicted in Figure 1 indicate that a pentacoordinate surface silicon is feasible. Thus polymerization would proceed so as to maximize the concentration of fluorides on the surface.

This surface polymerization scheme accounts for several observations regarding acidic fluoride catalyzed gels. They tend to be hydrophobic and water poor. It is difficult to retain large amounts of fluoride in silica glasses that are consolidated from sol gel derived powders or monoliths. During stabilization at high temperatures fluoride is lost as HF and SiF_3. [13]

REFERENCES

1. R.K.Iler, The Chemistry of Silica, Wiley, New York (1979).
2. R.J.P. Corriu, D. LeClerq, A. Vious, M. Pathe, and J. Phalippou, in Ultrastructure Processing of Ceramics, Glasses and Composites, L.L. Hench and D. R. Ulrich, Eds., Wiley, New York (1986).

3. E.M. Rabinovich, D.M. Krol, N.A. Kopylov and P.K. Gallagher, Journal of the American Ceramic Society, 72, 7 (1989), 1229.
4. E.J.A. Pope, and J.D. Mackenzie, Journal of Non-Crystalline Solids, 87 (1986), 185.
5. C.J. Brinker, and G.W. Scherer, Sol-Gel Science, The Physics and Chemistry of Sol-Gel Processing, Academic Press Inc., San Diego, Ca. (1990).
6. L.W Burggraf, L. P. Davis, and M. S. Gordon, in Ultrastructure Processing of Advanced Materials, D. R. Uhlman, and D.R. Ulrich, Eds., Wiley, New York (1992), 47-55.
7. L.P. Davis, and L.W. Burggraf, in Science of Ceramic Chemical Processing, L.L. Hench and D.R. Ulrich, Eds., Wiley, New York (1986), 400.
8. L.P. Davis, and L.W. Burggraf, in Ultrastructure Processing of Advanced Ceramics, J.D.Mackenzie, and D.R. Ulrich, Eds., Wiley, New York (1988),367.
9. S. Hayakawa and L.L. Hench, "Molecular Orbital Models of Silica Clusters Modified by Fluorine", Submitted to the Journal of Non-Crystalline Solids April (1998).
10. A.M. Bond, and G. T. Hefter, Critical Survey of Stability Constants and Related Thermodynamic Data of Fluoride Complexes in Aqueous Solution, IUPAC Chemical Series, 27, Pergamon Press, New York (1980).
11. K.W. Powers, The Development and Characterization of Sol Gel Substrates for Chemical and Optical Applications, Ph.D. Dissertation, University of Florida, Gainesville, FL (1998).
12. CRC Handbook of Chemistry and Physics, 50th Edition, Chemical Rubber Company, Cleveland, Oh., (1970).
13. K. Nassau, E.M. Rabinovich, A.E. Miller, and P.K. Gallagher, Journal of Non-Crystalline Solids, 82 (1986), 78.

THE SOL-GEL PROCESSING OF HYDROLYSATES DERIVED FROM CARBOXYLIC ACID-MODIFIED TITANIUM ISOPROPOXIDE

P.A. Venz[1], J.L. Woolfrey[2], J.R. Bartlett[2] , D.J. Cassidy[2] and R.L. Frost[1]
[1] Centre for Instrumental and Developmental Chemistry, Queensland University of Technology, 2 George Street, Brisbane, QLD, 4001, Australia.
[2] Materials Division, Australian Nuclear Science and Technology Organisation, Private Mail Bag 1, Menai, NSW, 2234, Australia.

ABSTRACT

The chemical modification of alkoxides has been used to control their reactivity during sol-gel processing, by affecting hydrolysis and condensation reactions. Little has been published on their effect on hydrolysate properties or subsequent sol-gel processing.

The composition, structure and surface properties of hydrolysates and sols produced from tetraisopropyltitanate modified with carboxylic acids ($C_nH_{2n+1}COOH$, where $n = 1$, 2 and 3) have been investigated by Raman spectroscopy, electrokinetic sonic analysis and titration. Hydrolysates initially formed at < 40 °C were amorphous oxy-hydroxides while those subsequently processed at higher temperatures crystallised to anatase. The rate of crystallisation at a given temperature decreased with increasing n.

The concentration of residual carboxylate species in the as-washed hydrolysates increased with increasing n. The relative abundance of terminal and bridging hydroxyls in the hydrolysates also varied with n, and the total concentration of surface hydroxyls increased with increasing n. Hydrolysates exhibited iso-electric points (IEP) at pH 5.9 to 7.3 for $n<3$. No IEP was observed when $n \geq 3$, and the corresponding hydrolysates exhibited a negative surface charge at pH > 2, attributed to sorbed carboxylate-species.

Such drastic changes in surface properties would affect subsequent processing methods, including sol-gel or colloidal processing, electrophoretic deposition, etc. The sorbed carboxylates are retained on the hydrolysate during heating to ~ 450 °C, affecting crystallisation and phase transformations during fabrication.

INTRODUCTION

The conditions generally used in colloidal sol-gel processing of concentrated solutions of alkoxides or inorganic salts initially produce oxide or hydroxide precipitates (hydrolysates) [1,2]. These hydrolysates may be calcined and used as powders or

To the extent authorized under the laws of the United States of America, all copyright interests in this publication are the property of The American Ceramic Society. Any duplication, reproduction, or republication of this publication or any part thereof, without the express written consent of The American Ceramic Society or fee paid to the Copyright Clearance Center, is prohibited.

de-aggregated by the addition of acid or base (peptisation), producing colloids (sol).

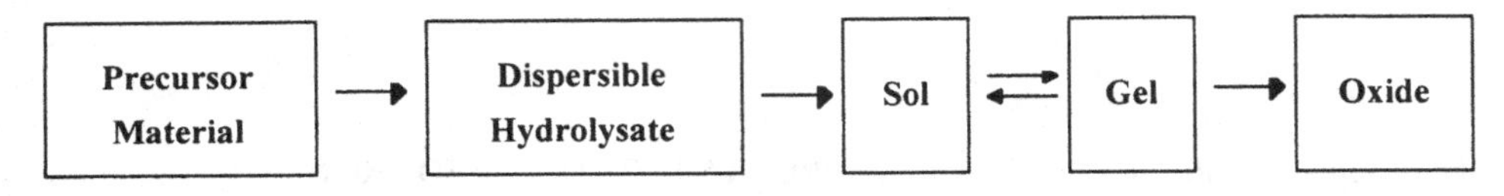

The sol may be further processed to produce gels, particles, coatings or monolithic solids.

Modification of alkoxides by nucleophilic substitution of chemical species into the coordination sphere of the metal atom (*e.g.* with carboxylic acids or β-diketones) has been used to control the reactivity of the alkoxide [3,4,5], affecting subsequent hydrolysis and condensation reactions. However, only limited results have been published on their effect on hydrolysate properties or subsequent processing [6,7].

The composition, structure and surface properties of hydrolysates and sols produced from tetraisopropyltitanate (TPT) modified with carboxylic acids ($C_nH_{2n+1}COOH$, where $n = 1$, 2 and 3) have been investigated by Raman spectroscopy, electrokinetic sonic analysis and titration. The affect of alkoxide modification on the structure and surface properties of the hydrolysates and sols produced, and on subsequent sol-gel processing, is discussed.

EXPERIMENTAL

Pure TPT and TPT/$C_nH_{2n+1}COOH$ mixtures (1:1 mole ratio, $n = 1,2$, and 3) were added to water at 298 K (50 moles H_2O per mole of TPT). The resulting hydrolysates were washed with water, and the pH of the slurries adjusted to ~2 by the addition of small quantities of 4 M HNO_3. The acidic slurries were analysed by electrokinetic sonic analysis (ESA, Matec MBS-800) and potentiometric titration (Radiometer Automatic Titrator) over the pH range from 2 to 12. The pH of the slurries was adjusted during titration by addition of 0.5 or 1.0 cm^3 aliquots of dilute NaOH, at five-minute intervals.

Raman spectra of the as-prepared and titrated hydrolysates were obtained with a Biorad FT-Raman II spectrometer, using a laser power of 500 mW, and an instrument resolution of 4 cm^{-1}. The dried hydrolysate powders were also characterised by XRD using a D500 Diffraktometer (Siemens), with Co K_α radiation.

Simultaneous thermogravimetric and differential thermal analysis (TGA/DTA) data (Setaram TAG 24) were obtained using a heating rate of 5 K min^{-1} in air.

RESULTS AND DISCUSSION

Surface Effects

XRD powder patterns of the as-washed hydrolysates showed only broad, diffuse peaks, suggesting that the hydrolysates were amorphous. In contrast, the corresponding Raman spectra exhibited bands at ~200, 440 and 570 cm^{-1}, which have been attributed to a highly-disordered structure with a small degree of short-range, "rutile-like" ordering [8].

Additional bands observed in the Raman spectra at 1416, 1455 and 1540 cm^{-1}, were assigned to residual sorbed carboxylate species with bidentate and chelating coordination geometries [8], Figure 1. The concentration of residual surface carboxylates which could not be removed by water washing, increased with increasing *n*.

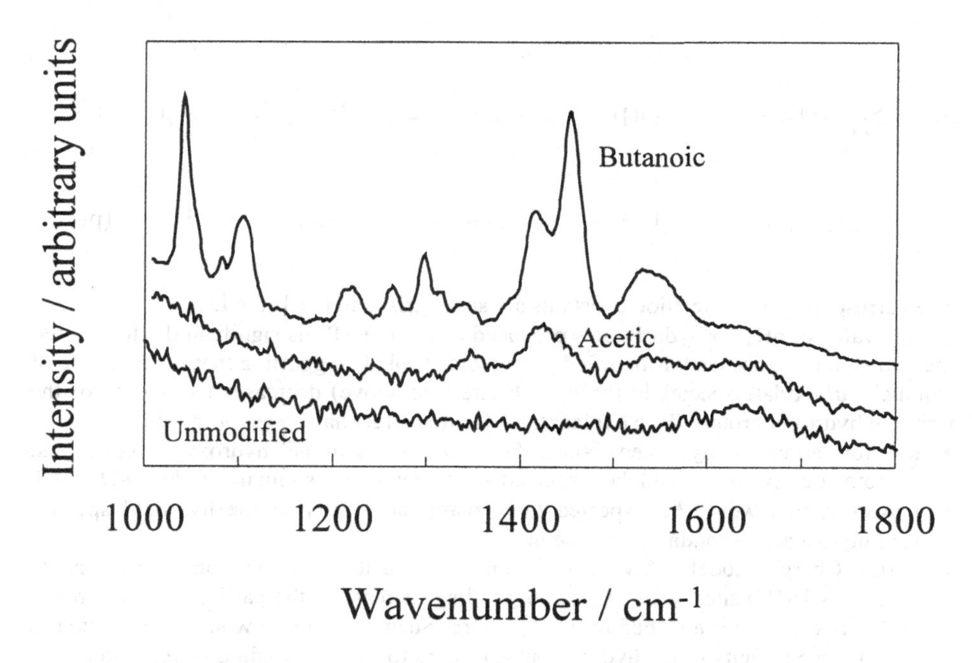

Figure 1. Raman spectra of residual carboxylate species in washed hydrolysates.

The electrokinetic properties of transition metal oxides are determined largely by the speciation of surface hydroxyl sites [9]. Such surface hydroxyls are amphoteric, and depending on the ambient pH, can act either as bases (generating positively-charged sites):

$$\geqslant Ti{-}OH \quad + H^+ \rightleftharpoons \quad \geqslant Ti{-}OH_2^+ \quad + H_2O$$

or acids (generating negatively-charged surface sites) [1,10,11,12,13]:

$$\geqslant Ti{-}OH \quad + [OH]^- \rightleftharpoons \quad \geqslant Ti{-}O^- \quad + H_2O$$

Titration of the hydrolysates with dilute NaOH revealed three acidic surface species [14]:

$$\geq Ti{-}OH_2^+ \quad + [OH]^- \rightleftharpoons \quad \geq Ti{-}OH \quad + H_2O \quad (pK_1)$$

$$\geq Ti{-}O(H){-}Ti\leq \quad + [OH]^- \rightleftharpoons \quad \geq Ti{-}O^-{-}Ti\leq \quad + H_2O \quad (pK_2)$$

$$\geq Ti{-}OH \quad + [OH]^- \rightleftharpoons \quad \geq Ti{-}O^- \quad + H_2O \quad (pK_3)$$

The corresponding dissociation constants are summarised in the Table I.

The value of pK_3 for hydrolysates produced from pure TPT is significantly different to the corresponding values for modified precursors, Table I, suggesting that the presence of residual carboxylate species in the hydrolysate (see above) decreases the acidity of the terminal hydroxyl groups. Two explanations for this effect have been proposed [14]:

- the formation of hydrogen bonds between the terminal hydroxyl species and carboxylate species would be expected to decrease the strength of the O-H bond. However, this would be expected to increase the acidity of the hydroxyl species, leading to a corresponding decrease in pK_3;
- Partial Charge Model (PCM) calculations indicate that coordination of carboxylate groups to Ti(IV) sites leads to a corresponding increase in the partial negative charge on hydroxyl species attached to the Ti centre. Such an increase would be expected to enhance the basicity of the hydroxyl sites leading to a corresponding increase in pK_3.

The experimental observations favour the latter explanation.

The total concentration of surface hydroxyls increased with increasing *n*, while the relative abundance of bridging and terminal hydroxyls also varied with *n*. Such profound changes in the surface speciation, associated with the presence of residual carboxylate species, had a significant affect on the electrokinetic properties of the hydrolysates. Typical variations in the electrophoretic mobility of the hydrolysates with pH are illustrated in Figure 2, while the IEP's of the hydrolysates are summarised in Table I. Modified hydrolysates with $n<3$ exhibited IEP's at pH 5.9 to 7.3. In contrast, no IEP was observed when $n\geq3$, with such hydrolysates exhibiting a negative surface charge at $pH \geq 2$, due to the retention of sorbed carboxylate species [14], which increased in abundance with increasing chain length, Figure 1. The retention of sorbed carboxylate species under acidic conditions can be explained using PCM theory:

- PCM calculations indicate that the partial charge on the carboxylate ligands decreases with increasing chain lengths, suggesting that the ligands should be more susceptible to protonation;

- However, protonated acetate and propanoate ligands both carry a positive partial charge ($\delta \sim +0.12$ and $+0.02$, respectively), suggesting that they are good leaving groups, readily desorbed in acidic solution, but with the latter desorbing at a much slower rate. In contrast, protonated butanoate ligands have a negative partial charge ($\delta \sim -0.07$), suggesting that they are poor leaving groups, and hence, not easily desorbed in acidic solution;

Thus, acetate ligands were completely desorbed after < 2.5 hours [15], but butanoate ligands were still evident at > 8 hours [16].

Table I. Surface Properties of titania hydrolysates.

Hydrolysate [1]	pK_1	pK_2	pK_3	IEP [2,3]	moles -OH per mole Ti		
					Bridging	Terminal	Total
Unmodified	2.9	6.3	8.6	5.9	0.08	0.20	0.28
Acetic	2.7	6.1	9.4	6.9	0.47	0.16	0.63
Propanoic	2.8	6.3	9.4	nd	0.43	0.11	0.54
Butanoic	2.7	6.0	9.4	nd	0.79	0.08	0.87

[1] refer to hydrolysates prepared from unmodified TPT, and from TPT modified with acetic, propanoic and butanoic acid, respectively.
[2] IEP, iso-electric point
[3] nd, no IEP was detected

Crystallisation

Crystallisation in Solution

Residual carboxylic acid in the amorphous, modified-precursors significantly influenced the rate at which changes occurred in the anatase bands of the corresponding hydrolysate Raman spectra during heating in water, or peptisation with HNO_3, at temperatures > 40 °C [15]. All systems exhibited a characteristic time, t_c, after which no further substantial changes were observed in component band position, intensity or the full-width at half-maximum of the band, Table II.

The value of t_c increased monotonically with the length of the carboxylic acid's hydrocarbon chain, with that of the TPT hydrolysate midway between the values for the corresponding acetic- and propanoic-modified systems. It was concluded that unreacted carboxylate species retained within the hydrolysate structure after hydrolysis and condensation have a significant influence on the rate of crystallisation of the hydrolysates.

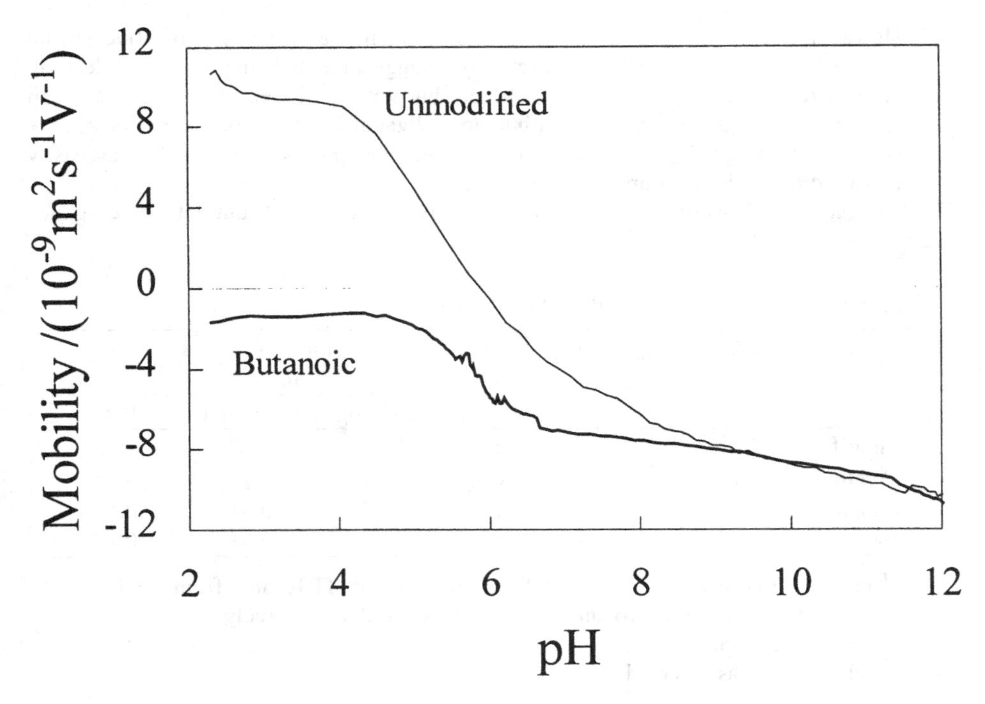

Figure 2. Electrokinetic properties of titania hydrolysates obtained from unmodified TPT and TPT modified with butanoic acid.

Table II. Effect of carboxylate chain length on the crystallisation of amorphous precursors at 60 °C

Precursor	t_c (minutes)	
	Samples peptised with HNO_3	Samples heated without HNO_3
Unmodified	180	≥360
Acetic	120	360
Propanoic	240	≥360
Butanoic	>420	>>420

t_c is the time to the end of crystallisation

Hydrolysate crystallisation can be viewed as the result of two key processes, both of which may be influenced by the presence of residual carboxylate species:

(a) A thermally-induced structural transformation: Crystallisation of the hydrolysates occurred very slowly at ambient temperature (requiring a period of several weeks), but at 60 °C it occurred within hours, regardless of whether or not HNO_3 was present [15].

(b) The condensation (olation, alcoxolation and/or oxolation) reactions involving hydroxyl groups in the hydrolysates [1,3]: Evidence that such reactions are involved in hydrolysate crystallisation is provided by the profound effect of acid catalysis on the rate of crystallisation during peptisation, when compared to systems heated in pure water, Table II. In the presence of HNO_3, hydroxyl groups on the hydrolysate surfaces are protonated by the acid, enhancing the rate of condensation [1,3] and, apparently, crystallisation. In addition, the carboxylic acid modifiers influenced the proportions in which the two predominant OH-species were present, Table I, and therefore the rate and extent of condensation during peptisation.

Crystallisation during calcination

TGA/DTA results showed that the sorbed carboxylates are retained on the hydrolysate and gels during heating to ~ 450 °C [16]. Hydrolysates were calcined in the TGA/DTA to 450 °C to remove any carboxylates, cooled to room temperature and then re-heated. The effect of initial carboxylate retention on the peak crystallisation and phase transformation (anatase → rutile) temperatures are summarised in Table III. The initial increases observed in the crystallisation and phase transformation temperatures for $n \leq 2$ is thought to be due to the retarding effects of residual organics/carbon on such processes [17,18]. The decreases observed in the butanoic-modified system are postulated to occur due to the large amounts of carboxylate initially retained in the hydrolysate, Figure 1, and the affect of the associated defects produced in the structure during subsequent burnout, accelerating the crystallisation and phase transformations [19,20]. Detailed results of this investigation will be reported elsewhere.

Molares *et. al.* [21] also found that the hydrolysis catalyst (acetic/oxalic acid) changed the crystallisation sequence and the anatase → rutile transformation temperature.

Table III. Effect of carboxylate retention on the peak crystallisation and phase transformation temperatures.

Precursor	Crystallisation (°C)	Phase Transformation (°C)
Unmodified	500-650	680
Acetic	636	691
Propanoic	640	691
Butanoic	567	665

Effect on Sol-Gel Processing

Such profound changes in surface properties would affect subsequent processing methods, such as sol-gel or colloidal processing, electrophoretic deposition, etc. In typical colloidal systems, the presence of an IEP means that the surface charge of the colloids can be manipulated by changing the pH:

- The strength of the charge can be controlled by the deviation from the IEP and is useful in sol-gel processing to enhance the rate of peptisation of the hydrolysate to form sols, using either acidic or basic solutions.
- An approach to the IEP can be used to destabilise and coagulate the colloidal system, promoting gelation and pore size control in the gel.
- Charge manipulation can also be used to promote deposition at a preferred electrode during electrophoretic deposition.

In contrast, in modified hydrolysates where $n \geq 3$, no IEP was observed and such hydrolysates exhibited a negative surface charge at $pH \geq 2$, due to sorbed carboxylate species. Such behaviour would:

- restrict the choices available during subsequent sol-gel processing;
- simplify processing requiring the deposition of a second colloid of opposite (+'ve) charge on the surface of the (-'ve charged) primary aggregates, since there would be a large pH range available to obtain the desired positive charge on the second colloid.

Retention of sorbed carboxylates during heating of the hydrolysate to ~ 450 °C affects the crystallisation sequence and phase transformations during calcination of powders at elevated temperatures, and thus the sintering sequence of such powders and gels.

CONCLUSIONS

The effect of alkoxide modification on the composition, structure and surface properties of the hydrolysates and sols produced from tetraisopropyltitanate (TPT) modified with carboxylic acids ($C_nH_{2n+1}COOH$, where $n = 1$, 2 and 3) has been investigated by Raman spectroscopy, electrokinetic sonic analysis and titration:

- Hydrolysates initially formed at < 40 °C were X-ray amorphous exhibiting a short-range order, with a "rutile like" structure, while those subsequently processed at higher temperatures crystallised to anatase.
- Residual carboxylate species were retained in the as-prepared hydrolysates. The concentration of surface carboxylates increased with increasing *n*.
- The relative abundances of terminal and bridging hydroxyls varied with *n*, and the total concentration of surface hydroxyls increased with increasing *n*.
- Hydrolysates obtained from TPT/$C_nH_{2n+1}COOH$ exhibited iso-electric points (IEP) at pH 5.9 to 7.3 for $n \leq 3$. In contrast, no IEP was observed when $n \geq 3$; such hydrolysates exhibited a negative surface charge at $pH \geq 2$, due to sorbed carboxylate species.
- The carboxylic acid species also influenced the rate and extent of condensation

during heating and peptisation of the hydrolysates, and the rate at which they crystallised to anatase.

- The sorbed carboxylates are retained on the hydrolysate during heating to ~ 450 °C and effect the crystallisation sequence and phase changes during calcination.

Such drastic changes in surface properties would affect subsequent processing methods, such as sol-gel or colloidal processing, electrophoretic deposition, *etc*. and calcination and sintering sequence of the powders and gels.

ACKNOWLEDGEMENTS

The Centre for Instrumental and Developmental Chemistry, the Research Centre of the School of Chemistry, of the Queensland University of Technology is gratefully acknowledged for financial support for this project. One of the authors (PAV) extends sincere thanks to The Australian Institute for Nuclear Science and Engineering (AINSE) for both financial and in-kind support.

REFERENCES

1. C.J. Brinker and G.W. Scherer, *Sol-Gel Science*, Academic Press, New York, 1990.
2. D. Segal, Chemical Synthesis of Advanced Ceramic Materials, Cambridge University Press, Cambridge, 1989.
3. J. Livage, M. Henry and C. Sanchez, Prog. Solid State Chem., 18 [4] 259 (1988).
4. S. Doeuff, M. Henry, C. Sanchez and J. Livage, J. Non-Cryst. Solids, 89 206 (1987).
5. M.J. Percy, J.R. Bartlett, J.L. Woolfrey, L. Spiccia and B.O. West, "The Influence of Acetylacetone During the Hydrolysis of Zirconium (IV) Propoxide", Proceedings of the Second International Symposium on Sol-Gel Science & Technology, Cairns. July, 1996, Trans. Amer. Ceram. Soc.,81 in press (1998).
6. S. Doeuff, M. Henry and C. Sanchez, Mat. Res. Bull. 25 1519 (1990).
7. S. Barboux-Doeuff and C. Sanchez, Mat. Res. Bull. 29 1 (1994).
8. P. A. Venz, R. L. Frost, J.R. Bartlett and J.L. Woolfrey, Spectrochim. Acta Part A, 53 969 (1997).
9. R.J. Hunter, Zeta Potential in Colloid Science - Principles and Applications, Academic Press, New York, 1981.
10. C. Contescu, V.T. Popa, J.B. Miller, E.I. Ko, and J.A. Schwarz, J. Catal. 157 244 (1995).
11. N. Spanos, I. Georgiadou and A. Lycourghiotis, J. Colloid Interface Sci. 172 374 (1995).
12. A. Mpandou and B. Siffert, J. Colloid Interface Sci. 102 138 (1984).
13. C. Ludwig and P.W. Schindler, J. Colloid Interface Sci. 169 284 (1995).
14. P.A. Venz, J.R. Bartlett, R.L. Frost and J.L. Woolfrey, "Electrokinetic Studies of Titania Derived from a Chemically-Modified Titanium Alkoxide", Proceedings of the Second International Symposium on Sol-Gel Science & Technology, Cairns. July, 1996, Trans. Amer. Ceram. Soc.,81 in press (1998).

15. P.A. Venz, J.R. Bartlett, R.L. Frost and J.L. Woolfrey, "Crystallisation of Titania Hydrolysates During Peptisation", Proceedings of the Second International Symposium on Sol-Gel Science & Technology, Cairns. July, 1996, Trans. Amer. Ceram. Soc.,81 in press (1998).
16. Unpublished results.
17. K. Terabe, K. Kato, H. Miyazaki, S. Yamaguchi, A. Imai and Y. Iguchi, J. Mater. Sci., 29 1617 (1994).
18. D.C. Hague and M.J. Mayo, J. Amer. Ceram. Soc., 77 [7] 1957 (1994).
19. C.N.R. Rao and K.J. Rao, "Phase Transformations in Solids", in Progress in Solid State Chemistry, 131-185 (1967).
20. G.A. Chadwick, "Metallography of Phase Transformations", Butterworths, London, p. 194 (1972).
21. B.A. Morales, O. Novaro, T. López, E. Sánchez and R. Gómez, J. Mater. Res., 10 [11] 2788 (1995).

SYNTHESIS OF MAGNESIA POWDERS FROM AN ALKOXIDE PRECURSOR

Mohan Menon, Julie L. Warren, and Jeffrey W. Bullard
Department of Materials Science and Engineering
University of Illinois at Urbana-Champaign
Urbana, Illinois 61801

ABSTRACT

Low-temperature routes are investigated for producing fine MgO powders from magnesium ethoxide precursors. Chemical and phase evolution of the precursor are tracked by differential scanning calorimetry (DSC), thermogravimetric analysis (TGA), x-ray powder diffraction (XRD), and by Diffuse Reflectance Infrared Fourier Transform Spectroscopy (DRIFTS). These data are reported as a function of temperature, heating rate, solution pH, and the type of free anions present in the precursor. We find that periclase MgO of high phase purity and high surface area can be formed in air at temperatures as low as 200°C.

INTRODUCTION

Finely divided, phase-pure crystalline MgO powder is of considerable technological importance, primarily because of the unique properties it imparts when used as a catalyst support. MgO can alter the reactivity of the metallic catalyst particles [1-3] and enhance the selectivity of the catalyst toward a given reaction. Several studies have also shown that MgO can stabilize alternate oxidation states in metals, and can inhibit both coarsening and evaporative mass loss of the metal constituent [4].

The most common way of producing MgO powders is by decomposition of either $MgCO_3$ or $Mg(OH)_2$. The particle size and specific surface area of powders obtained this way are strongly dependent on the particle size of the parent phase [5] and the annealing conditions under which the decomposition occurs. For instance, Kim *et al.* were able to obtain $\approx$ 2-nm cross-section MgO particles by decomposing 100–200 nm particles of $Mg(OH)_2$ at 275°C in a vacuum ($\approx 5 \times 10^{-5}$ Torr). However, under more common circumstances— air annealing of micron-sized particles, the resulting MgO has much lower specific surface area [1, 2].

To the extent authorized under the laws of the United States of America, all copyright interests in this publication are the property of The American Ceramic Society. Any duplication, reproduction, or republication of this publication or any part thereof, without the express written consent of The American Ceramic Society or fee paid to the Copyright Clearance Center, is prohibited.

Several investigators have examined the possibility of synthesizing MgO powders from liquid precursors. Lopez *et al.* [6] obtained high-surface area powders by hydrolysis and condensation of magnesium ethoxide, and examined the influence of solution pH by adding one of several hydrolysis catalysts (HCl, CH_3COOH, $H_2C_2O_4$, and NH_4OH). Building on that work, Bokhimi *et al.* used Rietveld refinement to determine the average particle size, lattice parameter, and Mg occupancy of powders prepared using each of these hydrolysis catalysts [7]. They reported that particle sizes < 100 nm could be obtained using any of the tested catalysts except HCl, which formed 150 nm particles. Because of intermediate formation of $C_2MgO_4 \cdot 2H_2O$ (Glushinskite) and $Mg(OH)_2$, most of the powders required an anneal at 600°C–900°C to initiate transformation to periclase MgO. However, these investigators found that, when acetic acid was used as the hydrolysis catalyst, Glushinskite and $Mg(OH)_2$ did not form and pure, 30-nm MgO powders* could be obtained at 150°C [7].

EXPERIMENTAL PROCEDURE

MgO precursor solution was prepared by a sol-gel route following a method similar to that of Bokhimi [7]. Magnesium ethoxide (Alfa Æsar, Ward Hill, MA) (MEO) was suspended in ethanol. The MEO was refluxed in acetic acid, which caused substitution of some of the ethoxy groups by acetyl groups. Hydrolysis was initiated at a pH of 3, 5.5, or 9 by adding one of several additional catalysts. The hydrolysis catalysts used were HNO_3 or HCl to achieve pH = 3, deionized water for pH = 5.5, and NH_4OH for pH = 9. The final precursor solution consisted of 0.5 mol/L of the chemically modified MEO. This solution is remarkably stable, showing no signs of degradation after several months of quiescent storage.

The MgO precursor solution was dried at 100°C to remove ethanol. The resulting white powder was analyzed by differential scanning calorimetry (DSC) and thermogravimetric analysis (TGA) (Netzsch STA 409, Germany), using a constant heating rate of 10°C/min and an Al_2O_3 reference powder to determine the temperatures at which organic pyrolysis and crystallization occurred.

Structural information about the powders at different temperatures was obtained by Diffuse Reflected Infrared Fourier Transform Spectroscopy (DRIFTS) (Nicolet 550, Madison, WI). Specimens were prepared by

*The particle size reported by these investigators was inferred from Rietveld refinement of a powder diffraction pattern, rather than by direct microscopic examination.

lightly crushing the powder and mixing it thoroughly with powdered KBr. Crystallinity and phase composition were investigated by X-ray θ-2θ diffraction (Rigaku D max, Japan). To determine the specific surface area, BET isotherms were obtained (Micromeritics ASAP 2400) by nitrogen adsorption from He carrier gas. All powders were outgassed at 150°C for 60 h prior to obtaining the isotherms.

RESULTS AND DISCUSSION

Fig. 1 tracks the weight loss from the precursors as a function of temperature, at a heating rate of 10°C/min. The precursors with pH 3 (HNO_3), pH 5.5 (H_2O), and pH 9 (NH_4OH) all exhibit some slight weight loss up to 200°C, which is probably due primarily to ethanol evaporation. At ≈360°C, all three of these precursors begin a period of rapid mass loss which is nearly complete by ≈400°C. Above 500°C, the mass of each of these three precursors is stable. The precursor with pH 3 (HCl) shows markedly different behavior: mass loss occurs continuously and gradually over the entire temperature range studied, and continues to lose mass up to 700°C.

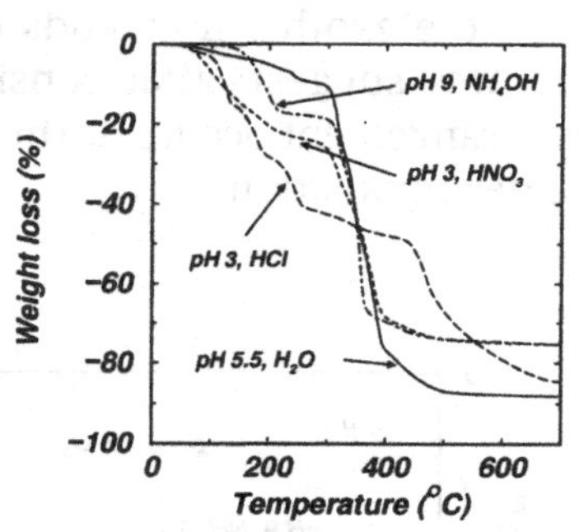

Figure 1: TGA curves for various precursors

Fig. 2 shows the powder diffraction patterns obtained for a powder prepared from the precursor at pH 5.5 and heated to 200°C and 360°C. The strong low-angle peaks (2θ =0°-20°) observed at 200°C correspond closely with β-magnesium acetate (triclinic $Mg(CH_3COO)_2$, which we will abbreviate as β-MgAc). Crystallization of β-MgAc from the amorphous precursor probably occurs just after evaporation of residual ethanol. Fig. 3 shows a plot of the DSC and TGA curves obtained on the bulk precursor by heating at 10°C/min. Evaporation of ethanol, which is responsible for the low-temperature endotherms and

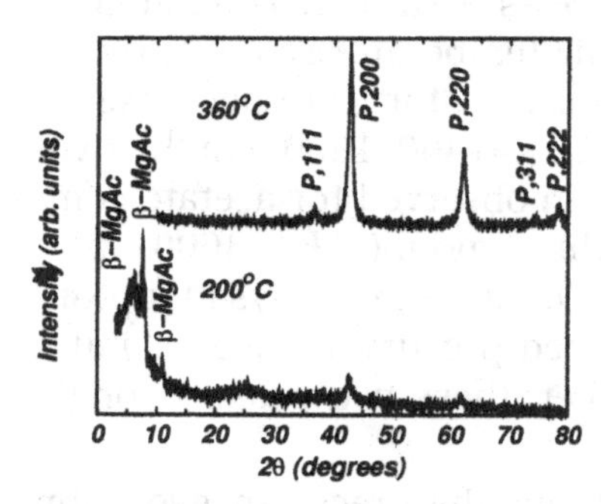

Figure 2: XRD pattern of pyrolyzed, pH 5.5 precursor

which results in ≈70% mass loss, appears to be complete by 100°C–120°C.

After ethanol evaporation, from 150°C–250°C, a broad exothermic signal suggests a structural transformation, since no mass loss is observed in the same range. The exothermic region is probably associated primarily with crystallization of β-MgAc. The fact that the exotherm extends over ten minutes suggests that considerable rearrangement occurs during pyrolytic decomposition.

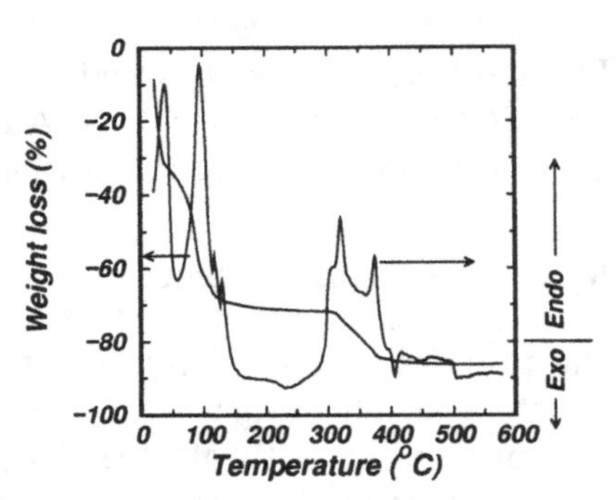

Figure 3: DSC/TGA of pH 5.5 precursor

Figs. 4 and 5 show DRIFTS spectra of precursors heated to 325°C. Three regions are of interest: (1) $k \approx 500\ cm^{-1}$; this region is associated with an Mg-O stretch mode, and an intense peak at this wavenumber are a signature of periclase MgO. Other spectra not shown here indicate that this peak grows with increasing annealing temperature as the precursor transforms to periclase.

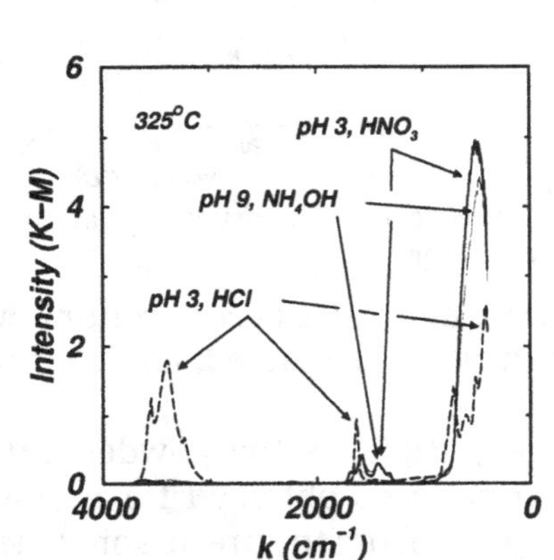

Figure 4: DRIFTS spectra at 325°C

(2) k=1400–1600 cm^{-1}; twin peaks centered about k=1500 cm^{-1} are often observed for acetate compounds. MgO also excites a minor peak in this region. (3) $k \approx 3600\ cm^{-1}$; only present in precursors prepared with HCl, this peak decays monotonically as temperature increases. HCl-catalyzed precursors also exhibit a unique peak at ≈1700 cm^{-1}. Taken together, these two peaks strongly suggest the presence of $Mg(OH)_2$.

Figs. 4 and 5 suggest several points about the precursor evolution as a function of its catalyst(s). HCl in combination with acetic acid as a catalyst appears to delay formation of MgO, and to perhaps favor early formation of $Mg(OH)_2$. The presence of $Mg(OH)_2$ could be definitively tested by x-ray diffraction on the HCl-catalyzed precursors; such data are unavailable at the time of this writing, but are planned in the very near future.

Among the other catalysts, both HNO_3 and NH_4OH, when combined with acetic acid, appear to promote phase evolution that is somewhat similar to that observed for the precursor having pH 5.5. In particular, there is no evidence from DRIFTS that $Mg(OH)_2$ forms, and it appears that acetate groups are present at 325°C. Finally, addition of either HNO_3 and NH_4OH to the precursor seem to enhance formation of MgO compared to precursors with only acetic acid added as a catalyst. Again, however, confirmation of the phases present awaits further x-ray diffraction studies.

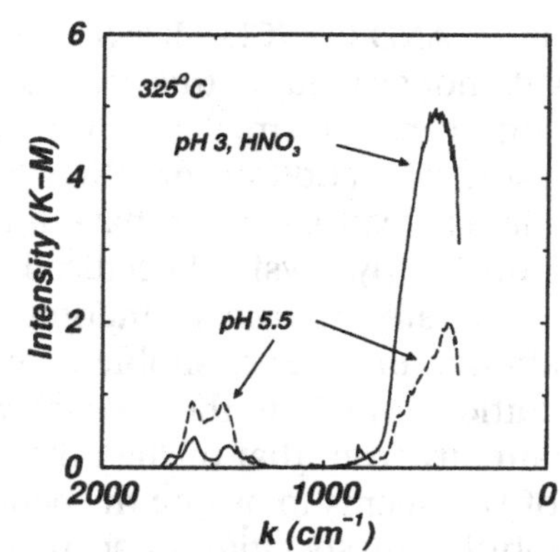

Figure 5: DRIFTS spectra at 325°C

Table I: Specific surface area of powders formed at 550°C using different precursor catalysts

Second Catalyst	Specific Surface Area (m^2/g)
HCl (pH 3)	4.2
HNO_3 (pH 3)	70.6
H_2O (pH 5.5)	80.2
NH_4OH (pH 9)	87.3

Finally, Table I gives the specific surface areas of powders obtained at 550°C as a function of the species added to the partially acetylated alkoxide precursor. Powders formed from precursors containing HCl have dramatically lower specific surface areas than those formed from the other precursors. Those other precursors appear to generate powders with fairly high specific surface areas, although not as high as reported by Bokhimi *et al.* [7], who used a slower heating rate and longer annealing times. Furthermore, increasing precursor pH appears to modestly increase the specific surface area of these powders. If one assumes the powders to have the theoretical density of periclase (3.58 g/cm^3), then the data reported in Table I correspond to equivalent spherical particle diameters of about 10 nm. Helium pyncometry measurements and electron microscopy observations, both planned for the near future, should reveal the quality and morphology of the particles.

SUMMARY AND CONCLUSIONS

As demonstrated by this work and by other investigators [6, 7], periclase MgO can be formed at <300°C from a partially acetylated alkoxide precursor. The addition of HNO_3 (pH 3) or NH_4OH (pH 9) as a second potential catalyst seems to have a modest influence on the phase evolution during pyrolysis. Specifically, both of these secondary catalysts may accelerate the transformation to periclase MgO while blocking the formation of other undesirable phases like $Mg(OH)_2$ or Glushinskite.

Additions of HCl to the precursor cause the phase evolution to differ substantially from that of the other precursors we examined. The presence of HCl seems to negate the beneficial effects of acetic acid, possibly by inhibiting dissociation of acetic acid into acetate radicals and hydronium ions. HCl seems to promote formation of $Mg(OH)_2$ and to form powders with low specific surface area.

Powders formed from the partially acetylated alkoxide with no other catalysts (pH 5.5) evolve through an intermediate β-magnesium acetate that forms prior to pyrolytic decomposition of the acetate groups. Although further x-ray diffraction studies will settle the issue, we expect from their DRIFTS spectra that precursors containing HNO_3 or NH_4OH as secondary catalysts also evolve through this same crystalline acetate precursor phase.

ACKNOWLEDGMENTS

This work was partially supported by the National Science Foundation under grant contract NSF DMR-9702610. We also thank Mr. John Bukowski for performing some of the DSC/TGA measurements.

REFERENCES

[1] S.R. Morris, R.B. Moyes, P.B. Wells, and R. Whyman, p. 247 in *Metal-Support and Metal-Additive Effects on Catalysis.* Edited by B. Imelik, *et al.*. Elsevier, Amsterdam, 1982.

[2] A.K. Datye and J. Schwank, "Fischer-Tropsch Synthesis on Bimetallic Ruthenium-Gold Catalysts," *J. Catal.* **93**, 256 (1985).

[3] N.K. Pande and A.T. Bell, "Influence of Support Composition on the Reduction of Nitric Oxide over Rhodium Catalysts," *J. Catal.* **98**, 7 (1986).

[4] J. Schwank, S. Galvano, and G.J. Parravano, "Isotopic Oxygen Exchange on Supported Ru and Au Catalysts," *J. Catal.* **63**, 415 (1980).

[5] M.G. Kim, U. Dahmen, and A.W. Searcy, "Structural Transformations in the Decomposition of $Mg(OH)_2$ and $MgCO_3$," *J. Am. Ceram. Soc.* **70**, [3] 146 (1987).

[6] T. Lopez, I. Garcia-Cruz, and R. Gomez, "Synthesis of Magnesium Oxide by the Sol-Gel Method: Effect of pH on the Surface Hydroxylation," *J. Catalysis* **127**, 75-85 (1991).

[7] Bokhimi, A. Morales, T. Lopez, and R. Gomez, "Crystalline Structure of MgO Prepared byt the Sol-Gel Technique with Different Hydrolysis Catalysts," *J. Solid State Chem.* **115**, 411-415 (1995).

SOL-GEL SYNTHESIS, SINTERING AND ELECTRICAL PROPERTIES OF NASICON HAVING NEW COMPOSITIONS, $Na_3Zr_{2-(x/4)}Si_{2-x}P_{1+x}O_{12}$

Enrico Traversa
Dipartimento di Scienze e Tecnologie Chimiche, Università di Roma "Tor Vergata", Via della Ricerca Scientifica, 00133 Roma, Italy

Laura Montanaro
Dipartimento di Scienza dei Materiali e Ingegneria Chimica, Politecnico di Torino, Corso Duca degli Abruzzi 24, 10129 Torino, Italy

Hiromichi Aono and Yoshihiko Sadaoka
Department of Materials Science and Engineering, Faculty of Engineering, Ehime University, Bunkyo-cho, Matsuyama 790-8577, Japan

ABSTRACT

Powders and pellets of new NASICON compositions have been synthesized using a mixed inorganic-organic sol-gel synthesis, by the preliminary formation of a pre-hydrolized TEOS xerogel. The investigated compositions can be described by the general formula $Na_3Zr_{2-(x/4)}Si_{2-x}P_{1+x}O_{12}$, obtained by keeping constant (= 3) the Na concentration at the optimum value reported for ionic conductivity, with x = 0 (the usual NASICON composition), 0.667, and 1.333. The xerogels were calcined at various temperatures in the range 400-800°C. The samples were analyzed by TG/DTA, BET measurements, XRD, EDS, and SEM. The powders were sintered into pellets at 1100°C. The sinterability increased with increasing the x value. This is attributed to the presence of a glassy phase and the occurrence of liquid phase sintering. The electrical conductivity of the NASICON sintered bodies, measured by a.c. impedance spectroscopy, is correlated with the lattice parameters of the hexagonal structure, and increased with increasing the x value.

INTRODUCTION

The discovery of the NASICON structure, $Na_{1+x}Zr_2Si_xP_{3-x}O_{12}$, represented a substantial improvement in solid electrolytes' development because its three-dimensional framework structure presents a high ionic conductivity, comparable to that of the two-dimensional networks, like β-alumina. One of the first investigations was performed by Hong [1] who demonstrated that solids in the compositional range $Na_{1+x}Zr_2Si_xP_{3-x}O_{12}$, with $0 < x < 3$, crystallize in the NASICON structure when heated to 1200°C. This structure has a rhombohedral symmetry, except in the interval $1.8 < x < 2.2$, where a small distortion to monoclinic symmetry occurs.

NASICON-type structures exist in a large triangular portion of the quaternary phase diagram Na_2O-ZrO_2-P_2O_5-SiO_2, delimited by the compositions: $NaZr_2P_3O_{12}$, $Na_4Zr_2Si_3O_{12}$, and $Na_4Zr_{1.25}P_2O_{12}$ [2]. Nevertheless, most of the studies have been concentrated on the above reported general formula [3-7], around

To the extent authorized under the laws of the United States of America, all copyright interests in this publication are the property of The American Ceramic Society. Any duplication, reproduction, or republication of this publication or any part thereof, without the express written consent of The American Ceramic Society or fee paid to the Copyright Clearance Center, is prohibited.

the composition x = 2. This is because the highest conductivity values have been observed in the compositional range $1.8 < x < 2.4$. Only few researches were dedicated to different compositions [2, 8-12]: Kohler *et al.* [2], and later Rudolf *et al.* [9,10] succeeded in preparing Zr-deficient monoclinic compositions. Outside the aforesaid triangular portion of the phase diagram, metastable NASICON phases were also synthesized [13].

In order to obtain zirconia-free materials, Von Alpen *et al.* [14] proposed a different general formula: $Na_{1+x}Zr_{2-x/3}Si_xP_{3-x}O_{12-2x/3}$. However, though zirconia-free, the materials with this formula consist of two phases [15]. Also the materials with the Hong compositions generally contain two (when zirconia-free) or three different phases. In fact, many Authors [12, 15, 16] have demonstrated the formation of a glassy phase (which is basically a sodium silicophosphate with some dissolved zirconia [16]) for many compositions, mostly in the silica-rich side; this phase has a high ionic conductivity, but it is rapidly degraded upon contact with water. In addition, the appearence of new, liquid phases was observed after high temperature treatments and correlated to an increase in powder sinterability [15-17].

It is thus clear that the synthesis of NASICON ceramics as pure phases is very difficult. Different wet chemical processes and sol-gel techniques [6, 18-20] have been investigated as an alternative to the synthesis by solid state reaction [1, 3, 5, 14]. The gel syntheses of NASICON ceramics have been also reviewed [21, 22]. Solution syntheses seem to offer a highly homogeneous material; the higher reactivity of the precursors could lead to purer phases and small grains with improved sinterability [12]. Generally, the crystallization of the amorphous precursors lead directly to the hexagonal structure of NASICON, using wet-chemical routes [5,6], whereas this structure is stable only at temperatures higher than 200-300°C when it is formed from the monoclinic structure yielded by the solid-state reactions [3]. It has been also pointed out, however, that the rhombohedral to monoclinic phase transition for NASICON is affected by the orientational disorder, in turn derived by the processing and thermal history of NASICON [23].

A commercial product (NGK Co. Ltd, Japan), having nominal composition $Na_3Zr_2Si_2PO_{12}$, was produced and investigated as CO_2 sensor [24-27]. However, the porosity of this materials was too high to ensure a sensing performance good enough for practical applications.

This paper deals with the preparation of NASICON materials with new compositions, for a further exploration of the above defined ternary field, having on its corners the compositions: $NaZr_2P_3O_{12}$, $Na_4Zr_2Si_3O_{12}$ and $Na_4Zr_{1.25}P_2O_{12}$. In particular, we selected compositions keeping the number of Na ions constant to three, in order to maintain constant the amount of Na ionic carriers present in $Na_3Zr_2Si_2PO_{12}$, but trying to improve the sinterability. The materials were prepared using a mixed inorganic-organic sol-gel synthesis, already used for the preparation of other complex systems [28, 29]. Since a direct relationship between conductivity and the size of the *c* parameter in the hexagonal cell has been clearly established [8], the outcoming modifications of the cell parameters as a function of the composition were studied and correlated with the electrical conductivity, the sinterability, and the microstructural evolution during the sintering of the materials tested.

EXPERIMENTAL PROCEDURE

NASICON-type materials in the $Na_3Zr_{2-(x/4)}Si_{2-x}P_{1+x}O_{12}$ system, with x = 0 (the usual NASICON composition, used as reference), 0.667, 1.333, and 2, were prepared. From now on they will be labelled samples A, B, C, and D, respectively.

The powders were prepared through an unconventional sol-gel technique by means of the preliminary formation of an amorphous solid (xerogel), according to the following procedure. A measured volume of tetraethylorthosilicate (TEOS) was added to ethanol and distilled water in order to have a TEOS/ethanol molar ratio of 0.1 and an ethanol/water molar ratio of about 2.5. These values were chosen considering the TEOS-water-ethanol ternary phase diagram [30], in order to perform the synthesis in the miscibile area of the diagram, and thus to better homogenize the reactants and try to limit the monoclinic zirconia segregation.

After an addition of HNO_3 1M up to a pH of about 0.5, the mixture was stirred for 30 minutes and then added, under stirring, with an aqueous solution of $NaNO_3$ and $ZrO(NO_3)_2$. About 30 minutes of stirring were needed to obtain a clear solution which was finally added with an aqueous solution of $(NH_4)_2HPO_4$: a gelatinous precipitate appeared, which after about half a hour of stirring was dried at 105°C in an oven and then calcined in air at 400°C for 30 minutes.

The powders calcined at 400°C were then ground in a planetary mill for 4 hours in absolute ethanol to obtain a narrow particle size distribution, lower than 5 μm, as controlled by laser granulometry. Each amorphous sample was submitted to simultaneous thermogravimetric and differential thermal analysis (TG-DTA, with a heating rate of 10°C/min in flowing air), in order to determine the crystallization temperature of the xerogels.

The phase evolution of the powders was analyzed by X-ray diffraction (XRD, CuKα radiation, λ = 1.54060 Å), after calcination at different temperatures. The XRD pattern were recorded at room temperature (20°C) with a Philips PW1710 diffractometer equipped with a graphite monochromator in the diffracted beam and a sample spinner (40 kV, 25 mA, 0.01 θ step, 5 s counting time). Before the measurements, the powders were ground in a planetary mill for 4 hours to obtain a particle size distribution lower than 5 μm. The cell parameters, the cell volume and the corresponding theoretical density were determined as a function of the composition, as reported in detail elsewhere [31]. The changes in the specific surface area after calcination at various temperatures were investigated by nitrogen adsorption. The morphological evolution of the powders as a function of composition and thermal treatments was observed by scanning electron microscope (SEM).

The powders calcined at 400°C were uniaxially (150 MPa) and then isostatically (200 MPa) pressed and sintered at 1100°C for 6 hours: slowly cooled (SC) and quenched (Q) samples were prepared to investigate the appearence of new phases and the presence of glassy phases during the high temperature sintering step.

For the electrical conductivity measurements, Au electrodes were deposited by sputtering on both sides of the NASICON pellets. The electrical conductivity of the pellets was determined by means of a.c. impedance spectroscopy measurements, performed at various temperatures in the frequency range 100 Hz to 10 MHz.

RESULTS AND DISCUSSION

The TG-DTA diagrams were characterized by a large weigth loss between 120° and 180°C associated to an endothermic peak, due to the loss of physisorbed water. A second endothermic peak associated to a weigth loss was observed between 550° and 650°C, due to the chemically bound water. The crystallization temperatures for all the prepared materials ranged between 650 and 700°C. Only the thermogram of sample D was not well resolved, showing two overlapping peaks.

The XRD patterns of the samples heated at the as-determined crystallization temperatures showed broad peaks of the NASICON structure. To achieve a better crystallization the powders were calcined at 900° and 1200°C for 30 minutes. All the products heated to 900°C showed broad peaks of the NASICON structure and were free of zirconia peaks. The compositions B and C, even after calcination at 1200°C, were free from foreign phases and showed hexagonal symmetry, while the sample A showed small peaks of tetragonal zirconia and presented a monoclinic structure. The sample D was formed by a mixture of two phases: $Na_2ZrP_2O_8$ [32], and a hexagonal NASICON phase. For this reason, we did not consider the use of the sample D for further investigations.

The XRD data are summarized in Table I. One can observe that a continuous reduction of the cell parameters was measured with decreasing the silicon content. After milling, all the powders presented a mean diameter of 5 μm and the upper limit of their particle size distribution did not exceed 20 μm.

Table II shows the values of the specific surface area (SSA) after calcination of the various powders at selected temperatures. At lower calcination temperatures the higher SSA values were measured for the samples richer in Si, added as the single alkoxide reactant (TEOS).

Table I. X-Ray Diffraction Data for Samples Crystallized at 1200°C for 30 minutes

Sample	Composition	Symmetry	Cell parameters (Å)
A	$Na_3Zr_2Si_2PO_{12}$	Monoclinic	$a_0 = 15.589$ $b_0 = 9.03$ $c_0 = 9.224$ $\beta = 124.09°$
B	$Na_3Zr_{1.833}Si_{1.333}P_{1.677}O_{12}$	Rhombohedral	$a_0 = 9.000$ $c_0 = 22.964$
C	$Na_3Zr_{1.667}Si_{0.667}P_{2.333}O_{12}$	Rhombohedral	$a_0 = 8.902$ $c_0 = 22.805$
D	$Na_3Zr_{1.5}P_3O_{12}$	Rhombohedral	$a_0 = 8.82$ $c_0 = 22.79$
	$Na_2ZrP_2O_8$	Card 35-0125	

Table II. Calcination Temperature Evolution of the Specific Surface Areas

Calcination Temperature (°C)	S.S. area (m^2/g)		
	sample A	sample B	sample C
400	148	120	71
500	20	56	49
600	19	44	43
800	1	1	12

However, the SSA strongly decreased for the Si-rich samples with increasing the temperature and approaching their crystallisation, while the SSA reduction was less strong for powder C.

SEM observations of the powders as-prepared and calcined at 800°C explained this behavior. At 400°C powder A was made of both dense grains and agglomerates made of submicronic particles. After calcination at 800°C larger dense domains of more than 20 μm in diameter were preferentially observed. This evolution was coupled to a change in the adsorption isotherm curves: at 400°C a type I isotherm was observed, typical of a microporous powder [33], while at 800°C the isotherm was of type II, characteristic for non porous materials, due to the sintering induced by calcination in this Si-rich composition. At 400°C powder B was also formed of both agglomerates and dense grains, but the amount of the agglomerates made of submicronic equiaxial particles was larger than in sample A. Calcination at 800°C induced a sligth grain growth and the solid presented a very small amount of microporosity, which justifies its lower SSA.

On the contrary, powder C was only made of small agglomerates of submicron particles at 400°C. With respect to the powders A and B, these grains underwent a reduced growth by calcining at 800°C. The appearence of micropores (type I isotherm) allowed the powder to preserve a larger SSA at higher temperatures.

Sintering at 1100°C confirmed the poor sinterability of the typical NASICON composition, as recently confirmed by other Authors [12]: composition A did not exceed a density of 70% of the theoretical value, even starting from high green density (about 55-65% of the theoretical value: 3.259 g/cm^3 for composition A, 3.196 for composition B). On the other hand, the density of the powders with B and C compositions were able to overcome 90% of the theoretical value. It must be emphasized also that we used a sintering temperature lower than the temperatures usually reported by other Authors [34]. Sintering at 1100°C was performed in order to control the loss of Na during the heat treatments. Moreover, all the samples underwent the same type of processing and thermal histories, and thus their different behavior should be attributed only to their different composition.

According to SEM observations, the low sinterability of powder A was linked to the presence of the above mentioned hard agglomerates, which generated a duplex structure with very dense and highly porous regions. No significant differences were observed between the slowly cooled and the quenched samples, after 6 hours soaking at 1100°C. The only clear evidence was a different amount of tetragonal zirconia after sintering: the XRD peaks were more evident in the slowly cooled sample, without further modifications of the patterns (Figure 1).

Tetragonal zirconia was present in a larger amount for sample B, especially for the SC sample (Figure 2). SEM observations of the samples B clearly showed the presence of a glassy phase at the high sintering temperature, supporting the hypothesis already suggested in the literature [12, 15, 16, 23] that NASICON undergoes a liquid phase sintering. This high temperature liquid or viscous phase gave rise to a glassy phase on cooling due to its high content of glass-forming oxides, such as silica and phoshorous oxide [15]. The appearence of this liquid phase may induce the segregation of zirconia crystals by a dissolution-precipitation process.

Traces of tetragonal zirconia were only present in the slowly cooled sample C, while after quenching the materials was made only of rhombohedral NASICON (Figure 3). If a liquid phase was present at the high sintering temperature, it would not be transformed in a glassy phase even in the quenched material, but it probably gave rise to the growth of preferentially elongated NASICON crystals, according to SEM observations, again through a dissolution-recrystallization mechanism.

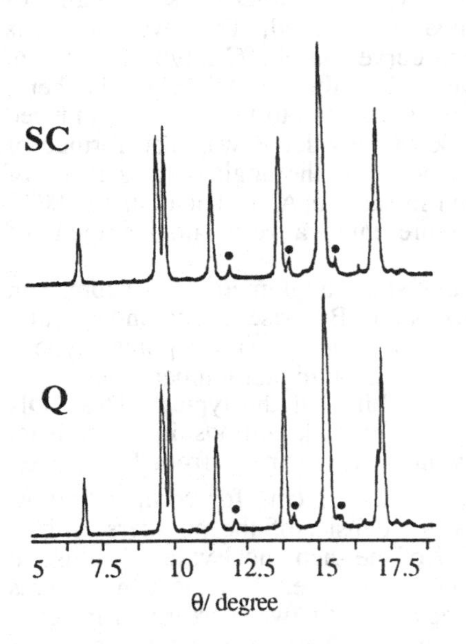

Figure 1. XRD patterns of the pellets, sintered at 1100°C, of powder A (SL =slow-cooled samples, Q = quenched samples, • = zirconia peaks).

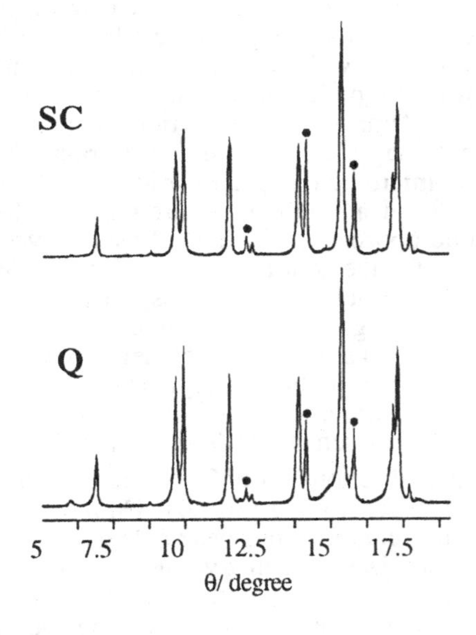

Figure 2. XRD patterns of the pellets, sintered at 1100°C, of powder B (SL =slow-cooled samples, Q = quenched samples, • = zirconia peaks).

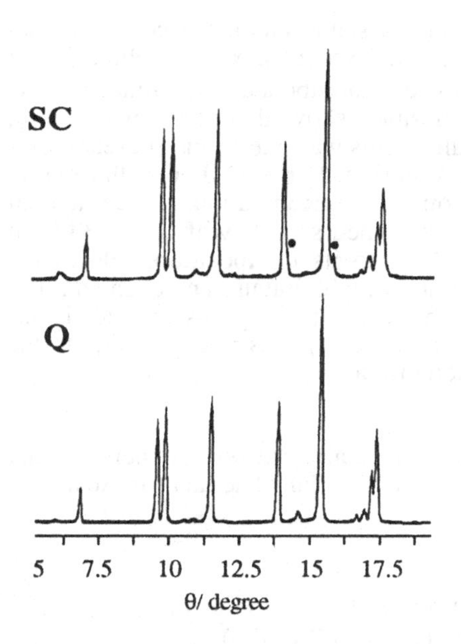

Figure 3. XRD patterns of the pellets, sintered at 1100°C, of powder C (SL =slow-cooled samples, Q = quenched samples, • = zirconia peaks).

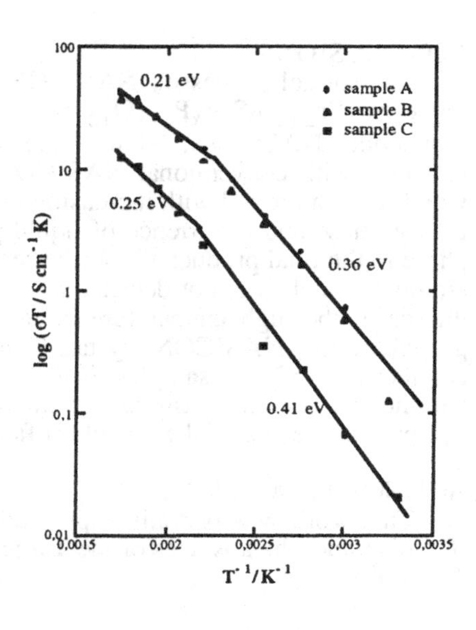

Figure 4. Temperature dependence of the conductivity measured by impedance spectroscopy for the pellets of NASICON tested.

Figure 4 shows the temperature-dependence of the conductivity for the three samples tested. The values of conductivity which are plotted in Fig. 4, derived from a.c. impedance spectroscopy measurements, correspond to the total conductivity of the bulk and the grain boundary. The calculated activation energies are in line with the values reported in the literature for NASICON. Although the sample A was very porous, its conductivity was the highest but close to the values measured for sample B. This can be interpreted as the balancing of two effects: the conductivity of sample B is increased by its higher sinterability, but is decreased by the reduction in the lattice constants, which means a reduction of the tunnel size for Na ions migration. Therefore, the samples A and B showed very close conductivities. On the other hand, for the sample C the conductivity was lower and the activation energy for ion migration was higher than those for the samples A and B. The smaller tunnel size strongly influenced the conductivity of the sample C. The values measured for conductivities and activation energies are in line with the values reported in the literature [23], although the sintering temperature was as low as 1100°C.

CONCLUSIONS

The sol-gel processed NASICON structure is stable along the compositional range $Na_3Zr_{2-(x/4)}Si_{2-x}P_{1+x}O_{12}$, with $0<x<1.333$. Except for x = 0, which is the monoclinic NASICON, all the samples showed rhombohedral symmetry. The structure with conventional NASICON composition showed a poor sinterability, which was improved with increasing the x value. This was due to the presence of a glassy phase and occurrence of liquid phase sintering. For x = 0.667, the glassy phase in the final product allows the segregation of significant amounts of tetragonal zirconia, which was not detected at all in the quenched sample with x = 1.333. In this case, the high temperature liquid phase, due to its composition, allows the growth of only NASICON crystals by a dissolution-reprecipitation mechanism. The conductivity of the samples was affected both by their porosity and lattice parameters. With increasing x, a reduction in conductivity was observed due to the decrease in the size of the tunnel for fast Na ion migration.

ACKNOWLEDGEMENTS

This work was partially supported by the National Research Council of Italy (CNR), under the auspices of the Targeted Project "Special Materials for Advanced Technologies II".

REFERENCES

1 H.Y-P. Hong, "Crystal Sructures and Crystal Chemistry in the System $Na_{1+x}Zr_2Si_xP_{3-x}O_{12}$," *Mater. Res. Bull.*, **11**, 173-82 (1976).

2 H. Kohler, H. Schulz, and O. Melhikov, "Composition and Conduction Mechanism of the Nasicon Structure. X-Ray Diffraction Study of Two Crystals at Different Temperatures," *Mater. Res. Bull.*, **14**, 1143-52 (1979).

3 J.P. Boilot, P. Salanié, G. Desplanches, and D. Le Potier, "Phase Transformation in $Na_{1+x}Zr_2Si_xP_{3-x}O_{12}$ Compounds," *Mater. Res. Bull.*, **14**, 1469-77 (1979).

4 D.H.H. Quon, T.A. Wheat, and W. Nesbitt, "Synthesis, Characterization and Fabrication of $Na_{1+x}Zr_2Si_xP_{3-x}O_{12}$," *Mater. Res. Bull.*, **15**, 1533-39 (1980).

5 G. Desplanches, M. Rigal, and A. Wicker, "Phase Transformation in an $Na_3Zr_2Si_xPO_{12}$ Ceramic," *Am. Ceram. Soc. Bull.*, **59**, 546-48 (1980).

6 B.E. Yoldas and I.K. Lloyd, "Nasicon Formation by Chemical Polymerization," *Mater. Res. Bull.*, **18**, 1171-77 (1983).

7 A. Caneiro, P. Fabry, H. Khireddine, and E. Siebert, "Performance Characteristic of Sodium Super Ionic Conductor Prepared by the Sol-Gel Route for Sodium Ion Sensors," *Analytical Chemistry*, **63**, 2550-57 (1991).

8 D. Tran Qui, J.J. Capponi, M. Gondrand, M. Saib, and J.C. Joubert, "Thermal Expansion of the Framework in Nasicon-Type Structure and its Relation to Na^+ Mobility," *Solid State Ionics*, **3/4**, 219-22 (1981).

9 P.R. Rudolf, M.A. Subramanian, and A. Clearfield, "The Crystal Structure of a Nonstoichiometric Nasicon," *Mater. Res. Bull.*, **20**, 643-51 (1985).

10 P.R. Rudolf, A. Clearfield, and J.D. Jorgensen, "A Time of Flight Neutron Powder Rietveld Refinement Study at Elevated Temperature on a Monoclinic Near-Stoichiometric Nasicon," *J. Solid State Chem.*, **72**, 100-12 (1988).

11 E. Breval and D. K. Agrawal, "Thermal Expansion Characteristics of NZP, $NaZr_2P_3O_{12}$-Type Materials: A Review," *Br. Ceram. Trans.*, **94**, 27-29 (1995).
12 N. Gasmi, N. Gharbi, H. Zarrhouk, P. Barboux, R. Morineau, and J. Livage, "Comparison of Different Synthesis Methods for Nasicon Ceramics," *J. Sol-Gel Sci. Technol.*, **4**, 231-37 (1995).
13 M. Barj, H. Perthuis, and Ph. Colomban, "Domaines d'existence, distortions structurales et modes de vibration des ions conducteur dans les reseaux hotes de type NASICON - Domain of Stability, Structural Distortions, and Vibrational Modes of Conducting Ions in the NASICON-type Host Lattices," *Solid State Ionics*, **11**, 157-77 (1983).
14 U. Von Alpen, M.F. Bell, and H.H. Hofer, "Compositional Dependence of the Electrochemical and Structural Parameters in the Nasicon Systems $Na_{1+x}Zr_2Si_xP_{3-x}O_{12}$," *Solid State Ionics*, **3/4**, 215-18 (1981).
15 A.K. Kuriakose, T.A. Wheat, A. Ahmad, and J. Dirocco, "Synthesis, Sintering and Microstructure of Nasicons," *J. Am. Ceram. Soc.*, **67**, 179-83 (1984).
16 A. Ahmad, T.A. Wheat, A.K. Kuriakose, J.D. Canaday, and A.G. McDonald, "Dependence of the Properties of Nasicons on Their Composition and Processing," *Solid State Ionics*, **24**, 89-97 (1987).
17 B.J. McEntire, G.R. Miller, and R.S. Gordon, "Sintering of Policrystalline Ionic Conductors: β-Alumina and Nasicon in Sintering Process"; pp. 517-524 in *Sintering Processes, Mater. Sci. Res.*, Vol. 13. Edited by G.C. Kuczynski. Plenum Press, New York, 1979.
18 H. Perthuis and Ph. Colomban, "Well Densified Nasicon Type Ceramics Elaborated Using Sol-Gel Process and Sintering at Low Temperatures," *Mater. Res. Bull.*, **19**, 621-31 (1984).
19 Ph. Colomban and J.P. Boilot, "Polymères inorganiques (xérogels et verres) dans les systèmes M_2O-$M'O_2$-SiO_2-P_2O_5-X_2O_3 - Inorganic Polymers (Xerogels and Glasses) in the Systems M_2O-$M'O_2$-SiO_2-P_2O_5-X_2O_3," *Rev. Chimie Minérale*, **22**, 235-55 (1985).
20 H. Perthuis and Ph. Colomban, "Sol-Gel Route Leading to Nasicon Ceramics," *Ceram. Intern.*, **12**, 39-52 (1986).
21 J.P. Boilot and Ph. Colomban, "Superionic Conductors from Sol-Gel Process"; pp. 304-29 in *Sol-gel Technology for Thin Films, Fibers, Preforms, Electronics and Specialty Shapes*. Edited by L. Klein. Noyes Publ., 1988.
22 Ph. Colomban, "Gel Technology in Ceramics, Glass-Ceramics and Ceramic-Ceramic Composites," *Ceram. Intern.*, **15**, 23-50 (1989).
23 Ph. Colomban, "Orientational Disorder, Glass/Crystal Transition and Superionic Conductivity in Nasicon," *Solid State Ionics*, **21**, 97-115 (1986).
24 N. Miura, S. Yao, Y. Shimizu, and N. Yamazoe, "High-Performance Solid-Electrolyte Carbon Dioxide Sensor with a Binary Carbonate Electrode," *Sensors and Actuators B*, **9**, 165-70 (1992).
25 S. Yao, Y. Shimizu, N. Miura, and N. Yamazoe, "Solid Electrolyte Carbon Dioxide Sensor Using Sodium-Ion Conductor and Li_2CO_3-$BaCO_3$ Electrode," *Jpn. J. Appl. Phys.*, **31** (part 2, n. 2B), 197-99 (1992).

26 N. Miura, S. Yao, Y. Shimizu, and N. Yamazoe, "Carbon Dioxide Sensor Using Sodium Ion Conductor and Binary Carbonate Auxiliary Electrode," *J. Electrochem. Soc.*, **139**, 1384-88 (1992).
27 Y. Sadaoka, Y. Sakai, M. Matsumoto, and T. Manabe, "Solid-State Electrochemical CO_2 Gas Sensors Based on Sodium Ionic Conductors," *J. Mater. Sci.*, **28**, 5783-92 (1993).
28 D. Mazza, M. Lucco Borlera, G. Busca, and A. Delmastro, "High Quartz Solid Solution Phases from Xerogels with Composition $2MgO.2Al_2O_3.5SiO_2$ (μ-Cordierite) and $Li_2O.Al_2O_3.nSiO_2$ (n = 2 to 4) (β-Eucryptite): Characterization by XRD, FTIR and Surface Measurements," *J. Europ. Ceram. Soc.*, **11**, 299-308 (1993).
29 D. Mazza and M. Lucco Borlera, "Effect of the Substitution of Boron for Aluminium in the β-Eucryptite $LiAlSiO_4$ Structure," *J. Europ. Ceram. Soc.*, **13**, 61-65 (1994).
30 C.J. Brinker and G.W. Scherer; p.109 in *Sol-gel Science*. Academic Press, San Diego, 1990.
31 M. Lucco Borlera, D. Mazza, L. Montanaro, A. Negro, and S. Ronchetti, "X-Ray Characterization of the New Nasicon Compositions $Na_3Zr_{2-x/4}Si_{2-x}P_{1+x}O_{12}$ with x = 0.333; 0.667; 1.000; 1.333; 1.667," *Powder Diffraction*, **12** [3], 171-74 (1997).
32 A. Clearfield, P. Jirustithipong, R.N. Cotman, and S.P. Pack, "Synthesis of Sodium Dizirconium Triphosphate from α-Zirconium Phosphate," *Mater. Res. Bull.*, **15**, 1603-10 (1980).
33 S. Lowell and J.E. Shield, *Powder Surface Area and Porosity*, Chapman and Hall, London, 1984.
34 J.P. Boilot, G. Collin, and Ph. Colomban, "Nasicon and Related Compounds: A Review"; pp. 91-122 in *Progress in Solid Electrolytes*. Edited by T.A. Wheat, A. Ahmad, and A.K. Kuriakose. ASCOR-CANMET, Ottawa, 1983.

PATTERNING OF INORGANIC-ORGANIC HYBRID FILMS USING CHEMICALLY MODIFIED METAL ALKOXIDE

Noriko Yamada, Ikuko Yoshinaga and Shingo Katayama
Advanced Technology Research Laboratories, Nippon Steel Corporation
3-35-1 Ida, Nakahara-ku, Kawasaki 211-0035, Japan

ABSTRACT

Optical properties and patterning of the inorganic-organic hybrid films prepared from diethoxydimethylsilane (DEDMS) and $Ti(OC_2H_5)_4$ modified with ethyl acetoacetate (EAcAc) were studied. The films showed the refractive index of 1.62-1.67 and small in-plane scattering. FT-IR study revealed that the chelate complex of $Ti(OC_2H_5)_4$ and EAcAc in the hybrid film were decomposed by UV-irradiation at 254 nm. The decomposition of the chelate complex brought about the structure change in the hybrids, leading to the solubility difference between UV-irradiated and unirradiated regions. This finding enabled us to pattern the hybrid films.

INTRODUCTION

The sol-gel process is based mainly on inorganic polymerization reactions of molecular precursors. Starting from metal alkoxides, an inorganic oxide network is obtained *via* hydrolysis-condensation reactions. The control of hydrolysis and condensation reactions of metal alkoxides is a key point in the sol-gel process because most metal alkoxides are very reactive and liable to give a precipitate. The chemical modification of metal alkoxides with β-diketones or β-ketoesters is effective to suppress the high reactivity [1-3]. Thus, a variety of oxide thin films have been prepared in a relatively convenient ways using chemically modified metal alkoxides [4-7]. Besides facilitating the fabrication of oxide films, chemical modification of metal alkoxides with β-diketones or β-ketoesters has been known to give photosensitivity to the gel film, which is applicable to the fine patterning of oxide thin films [8-11].

To the extent authorized under the laws of the United States of America, all copyright interests in this publication are the property of The American Ceramic Society. Any duplication, reproduction, or republication of this publication or any part thereof, without the express written consent of The American Ceramic Society or fee paid to the Copyright Clearance Center, is prohibited.

A low-temperature route through the sol-gel process enables the incorporation of organic molecules or dyes into an oxide gel matrix. However, the motion of organic molecules such as isomerization is restricted due to the rigidity of inorganic oxide network structure [12]. In addition, the fabrication of oxide thick films by the sol-gel method is difficult because of stress-induced cracking caused by the shrinkage during drying. Inorganic-organic hybrids prepared by the sol-gel process are new materials which can combine properties of both inorganic and organic materials [13-15]. The organic groups introduced into an inorganic network provide softness which makes the fabrication of thick and flexible films possible. These materials are promising in optical devices, photonics, sensors, catalysis and other fields. The authors have applied chemical modification of reactive metal alkoxides with β-diketones in the synthesis of inorganic-organic hybrids containing various inorganic components [16, 17].

The aim of the present work is to investigate the optical properties and patterning of inorganic-organic hybrid films using metal alkoxides modified with β-ketoester.

EXPERIMENTAL

The raw materials used were diethoxydimethylsilane (DEDMS, Tokyo Kasei Kogyo Co.), $Ti(OC_2H_5)_4$ (Tokyo Kasei Kogyo Co.), ethyl acetoacetate (EAcAc, Kanto Chemical Co.) and silanol-terminated polydimethylsiloxane (PDMS, average molecular weight = 800, Shinetsu Chemical Co.). $Ti(OC_2H_5)_4$ and EAcAc in a 1 : 2 molar ratio were mixed in 2-ethoxyethanol under stirring followed by the addition of DEDMS. Hydrolysis was carried out with acetic acid as a catalyst. The molar ratio of $Ti(OC_2H_5)_4$: DEDMS : acetic acid : H_2O was 4 : 1 : 0.25 : 5. Then 10 ml of the above solution and 0.5 ml of PDMS were mixed together and served as a coating solution. Films were prepared on Si or glass substrates by spin coating at a rotation speed of 3000 rpm. A drying was performed in an oven at 70℃ for 5 min. The substrates were then heat-treated at 150 - 400℃ for 10 min in air.

The film thickness was obtained using a profilometer. An ellipsometer equipped with a He-Ne laser was used to measure the refractive index. For patterning hybrid films, UV-light at 254nm with a power of 7 mW / cm^2 was irradiated on the films. Irradiation time was 5 min. The films before and after UV-irradiation were characterized by FT-IR.

RESULTS AND DISCUSSION

Change of Film Thickness and Refractive Index by Heat-Treatment

Figures 1 and 2 respectively show plots of the film thickness and the

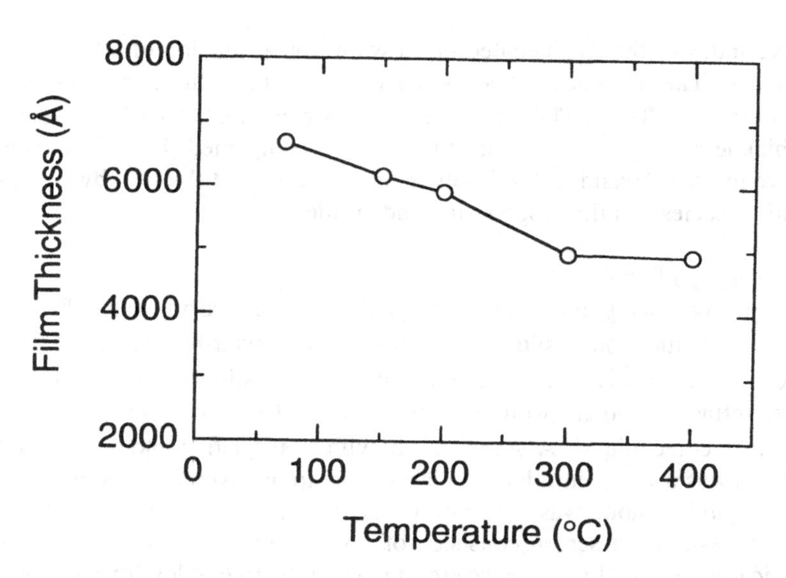

Figure 1. Dependence of film thickness on heat-treatment temperature.

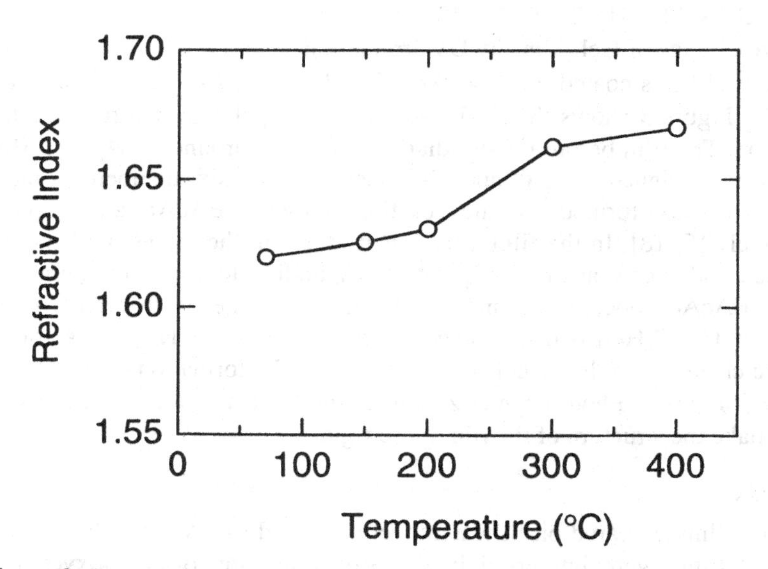

Figure 2. Dependence of refractive index on heat-treatment temperature.

refractive index of the films coated on Si wafers as a function of a heat-treatment temperature. The thickness steadily decreased with increasing heat-treatment temperature until 300℃. The refractive index became higher with the decrease in film thickness. The decrease in thickness accompanied by the increase in refractive index is considered to imply the densification of the film by the removal of volatile species and the progress in condensation.

Optical Waveguiding

A planer waveguide of the inorganic-organic hybrid was fabricated by coating the solution on a slide glass substrate followed by drying at 70℃ and heat-treating at 300℃. The refractive index of the slide glass was 1.520. The effective refractive indices were measured for the TM mode using a He-Ne laser by the *m*-line technique. A glass prism with a refractive index of 1.878 (λ =632.8nm) was used to introduce the laser beam to the hybrid waveguide.

One guided mode was observed in the planer waveguide. The output beam pattern observed on a screen consisted of a spot and a very week *m*-line, which was indicative of small in-plane scattering. The effective index for the TM_0 mode was 1.61.

PATTERNING OF HYBRID FILMS

Structure Change of Gel Films by UV-Irradiation

Hybrid films coated on Si wafers after drying at 70℃ were irradiated with UV-light. Figure 3 shows the FT-IR spectra of the gel film before and after UV-irradiation. The film before UV-irradiation had peaks around 1534 and 1618 cm^{-1}, which were assigned to C-C and C-O vibrations of six-membered ring of the chelate complex formed by the reaction between EAcAc and $Ti(OC_2H_5)_4$, respectively [5, 18]. In the film after UV-irradiation, the intensity of these peaks decreased and a new absorption peak corresponding to C=O stretching for keto type of EAcAc appeared around 1730 cm^{-1}. This means that part of EAcAc bonded to $Ti(OC_2H_5)_4$ is dissociated to form free EAcAc. The release of EAcAc from the chelate complex is considered to lead to the formation of new bonds, Ti-O-Ti and Ti-O-Si. Photopolymerization of the hybrid through such new bonds would make the structure of the film more rigid.

Patterning

Gel films prepared on Si wafers were exposed to UV-light through a mask. Irradiated films were immersed in the mixed solution of 2-ethoxyethanol and $0.5mol \cdot dm^{-3}$ HCl in a volumetric ratio of 1:1. They were subsequently rinsed in pure water. While the unirradiated area was soluble in the mixed solution, the

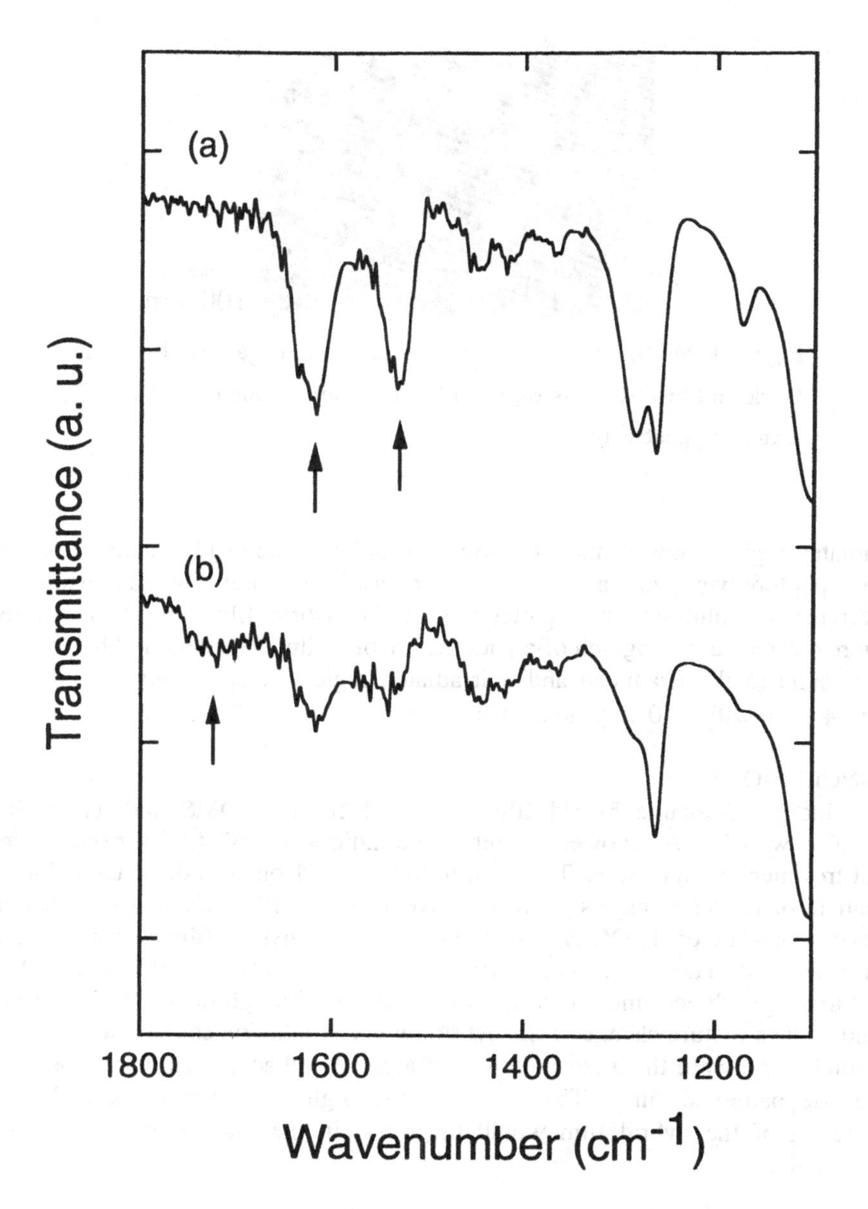

Figure 3. FT-IR spectra of the hybrid film (a) before and (b) after UV-irradiation.

Figure 4. Micrograph of a patterned inorganic-organic hybrid film. Dark and bright areas represent inorganic-organic hybrid and Si wafer, respectively.

irradiated region showed markedly lower solubility. The rigid structure resulting from photopolymerization seems to decrease the solubility. By using the difference in solubility, fine patterning of the hybrid film was demonstrated. Figure 4 shows a micrograph of a patterned hybrid film. The dark and bright parts correspond to the irradiated and unirradiated regions, respectively. The periods was 24 μ m with a 50/50 ratio of stripe to gap.

CONCLUSIONS

Inorganic-organic hybrid films prepared from DEDMS and $Ti(OC_2H_5)_4$ modified with EAcAc showed the refractive indices of 1.62-1.67 depending on a heat-treatment temperature. The hybrid film coated on a slide glass exhibited small in-plane scattering as a planer waveguide. FT-IR study revealed that the chelate complex of $Ti(OC_2H_5)_4$ and EAcAc in the hybrid film decomposed by UV-irradiation. The dissociation of EAcAc from the chelate complex is thought to lead to a rigid three-dimensional network structure through the formation of new bonds. The structure change of the hybrid by UV-irradiation caused the difference in solubility toward the mixed solution of alcohol and acid, making it possible to fabricate patterned films. The patternability, high refractive index and small scattering of the hybrid film would be useful in the fabrication of a channel waveguide.

ACKNOWLEDGEMENT

This work was supported by NEDO as part of the Synergy Ceramics Project under the Industrial Science and Technology Frontier (ISTF) Program promoted by AIST, MITI, Japan. The authors are members of the Joint Research Consortium of Synergy Ceramics.

REFERENCES

1. J. Livage, M. Henry and C. Sanchez, "Sol-gel chemistry of transition metal oxides," *Prog. Solid State Chem.*, **18**, 259-341 (1988).
2. D. C. Bradley, R. C. Mehrotra and D. P. Gaur, "Reactions with β-diketones and β-ketoesters"; pp. 209-217 in *Metal Alkoxides*, Edited by D. C. Bradely, R. C. Mehrotra and D. P. Gaur. Academic Press, London, 1978.
3. C. Sanchez, J. Livage, M. Henry and F. Babonneau, "Chemical modification of alkoxide precursors," *J. Non-Cryst. Solids*, **100**, 65-76 (1988).
4. K. Yamada, T. Y. Chow, T. Horihata and M. Nagata, "A low temperature synthesis of zirconium oxide coating using chelating agents," *J. Non-Cryst. Solids,* **100**, 316-320 (1988).
5. H. Uchihashi, N. Tohge and T. Minami, "Preparation of amorphous Al_2O_3 thin films from stabilized Al-alkoxides by the sol-gel method," *J. Ceram. Soc. Jpn.*, **97**, 396-399 (1989).
6. R. Nass and H. Schmidt, "Synthesis of aluminum sec-butoxide modified with ethylacetoacetate: An attractive precursor for the sol-gel synthesis of ceramics,", *J. Non-Cryst. Solids*, **121**, 329-333 (1990).
7. N. Tohge, S. Takahashi and T. Minami, "Preparation of PnZaO3-PbTiO3 ferroelectric thin films by the sol-gel process," *J. Am. Ceram. Soc.*, **74**, 67-71 (1991).
8. K. Shinmou, N. Tohge and T. Minami, "Fine-patterning of ZrO2 thin films by the photolysis of chemically modified gel films," *Jpn. J. Appl. Phys.*, **33**, L1181-1184 (1994).
9. N. Tohge, K, Shinmou and T. Minami, "Effects of UV-irradiation on the formation of oxide thin films from chemically modified metal-alkoxides," *J. Sol-Gel Sci. Technol.*, **2**, 581-585 (1994).
10. N. Tohge, K. Shinmou and T. Minami, "Photolysis of organically modified gel films and its application to the fine-patterning of oxide thin films," *SPIE Proc. Sol-Gel Optics III*, **2288**, 589-598 (1994).
11. N. Tohge and G. Zhao, "Formation of photosensitive alumina-based gel films and their application to the fine-patterning," *SPIE Proc. Sol-Gel Optics IV*, **3136**, 176-186 (1997).
12. D. Levy, S. Einhorn and D. Avnir, "Application of the sol-gel process for the preparation of photochromic information-recording materials: synthesis, properties, mechanisms," *J. Non-Cryst. Solids*, **113**, 137-145 (1989).

13. G. Philipp and H. Schmidt, "New Materials for Contact Lenses Prepared from Si- and Ti-alkoxides by the Sol-Gel Process," *J. Non-Cryst. Solids*, **63**, 283-292 (1984).
14. H. Schmidt, "New Type of Non-crystalline Solids between Inorganic and Organic Materials," *J. Non-Cryst. Solids*, **73**, 681-691 (1985).
15. G. L. Wilkes, B. Orler and H. Huang, "Ceramers: Hybrid Materials Incorporating Polymeric/Oligomeric Species into Inorganic Glasses Utilizing a Sol-Gel Approach," *Polym. Prep.*, **26**, 300-302 (1985).
16. S. Katayama, I. Yoshinaga and N. Yamada, "Synthesis of Inorganic-organic hybrids from metal alkoxides and silanol-terminated polydimethylsiloxane"; pp. 321-326 in *Better Ceramics Through Chemistry VII*, Edited by B. K. Coltrain, C. Sanchez, D. W. Schaefer and G. L. Wilkes. *Mat. Res. Soc. Proc.*, **435**, Pittsburgh, PA, 1996.
17. N. Yamada, I. Yoshinaga and S. Katayama, "Synthesis and Dynamic Mechanical Behaviour of Inorganic-Organic Hybrids Containing Various Inorganic Components," *J. Mater. Chem.*, **7**, 1491-1495 (1997).

TAILORING OF THE NANO/MICROSTRUCTURE OF HETEROGENEOUS CERAMICS BY SOL-GEL ROUTES

Philippe Colomban*
LADIR, CNRS
2, rue Henri Dunant
94320 Thiais, France

ABSTRACT

The optimization of the various and sometimes conflicting properties of advanced ceramics implies the achievement of a specific microstructure. This paper discuss the interest of sol-gel methods to tailor the micro- and nanostructure of multiphase and functionally graded ceramic-ceramic and metal-ceramic composites for thermostructural and/or microwave applications.

INTRODUCTION

One of the key steps in the advancement of the ceramic processing is the achievement of a special and controlled microstructure. This is particularly true for composites : **i)** the dissipative fracture behavior results from the presence of a very compliant interphase between the reinforcement and the matrix, **ii)** the high mechanical strength of a 3D reinforced composite results from both the preservation of the pristine state of the reinforcement, and a net-shape, crack-free, sintering, preserving the continuity of the matrix, **iii)** the high microwave absorption of metal-ceramic composites requires a homogeneous dispersion of nanosized metal precipitates within a dielectric matrix, etc. The objective of this review, based on the works conducted at the ONERA, is to show that the sol-gel route which has been mainly used to prepare homogeneous materials[1-4] constitutes a unique tool for the processing of multiphase materials, with tailored micro- and

*Also at DMSC, ONERA, BP 72, 92322 Chatillon France

To the extent authorized under the laws of the United States of America, all copyright interests in this publication are the property of The American Ceramic Society. Any duplication, reproduction, or republication of this publication or any part thereof, without the express written consent of The American Ceramic Society or fee paid to the Copyright Clearance Center, is prohibited.

nanostructure, given the unusual characteristics of gels and xerogels, such as the low viscosity of gels, the gelation, the meso/microporosity of the polymerized oxide network, and the easy nucleation of one phase from a pristine amorphous xerogel. This project takes inspiration in the "old" whiteware technology based on the natural colloids properties which makes it possible to achieve enamel coating, cold molding, porosity control and use, etc.

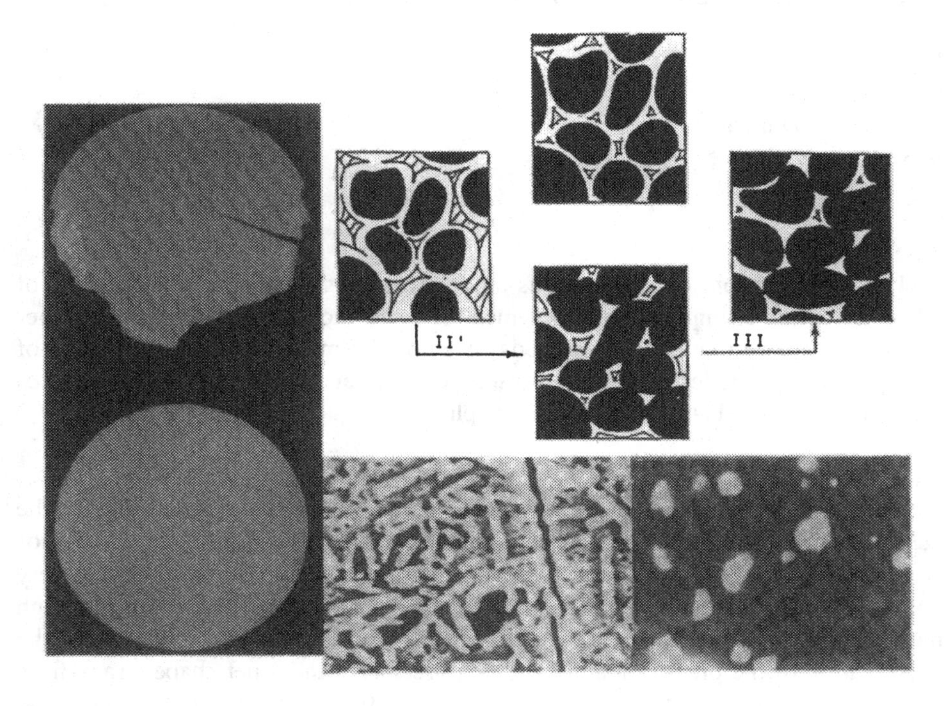

Fig. 1 : The use of the gel viscosity. *Top* : gel embedding (in white) and compaction of particulate reinforcements/grains (in black), porosity is dashed ; after compaction, gel is squeezed between the hard particles. *Left* : 1600°C sintered pellets, made of 30 wt% Al_2O_3 platelets in alumina matrix, with (bottom) or without (top) addition of a ZrO_2 gel (diameter : 30mm). *Bottom* : micrographs of polished sections of mullite matrix with 30 wt% platelets (left, pores appear in black) and plus addition of ZrO_2 gel (right, zirconia grains appear in white, mean size : 1 μm).

COLD MOLDING AND GREEN DENSIFICATION OF GEL CONTAINING BODIES : THE USE OF THE GEL VISCOSITY

To date, powders have mainly been coated in order to modify their processability (easy dispersion, no flocculation, etc.) and their sintering[2,5]. Sol-gel embedding has also been used to modify the electrical properties of the final ceramics, e.g. by achievement of an intergranular second phase[2], and to improve the dispersion of short fibers[6]. The homogeneous dispersion of anisotropic particles in a powder and the compaction of the resulting mixture become difficult to achieve when the content of added particulate reinforcements (whiskers, platelets, short fibers) reaches values (ca 10-30%) increasing the toughness significantly. The use of liquid aids (colloid processing) counteracts this drawback, although imperfectly[7]. Gel embedding offers a new route for the mixing of powders and particulate reinforcements thanks to the incorporation of a viscous substance which reactivity can be tailored by composition design[8]. Cold molding makes it possible to prepare shaped and crack free pieces (Fig. 1, left). The process is the following :

Table I : R.T. flexural strength (σ_{RT}), porosity (P) and toughness (K_{IC}) of sintered pellets made with mullite and alumina powders, with or without platelet reinforcement (Some data are reprinted from Composites Science and Technology, **58**, J.L. Lagrange and Ph. Colomban, "Double Particle Reinforcement of CMC's Prepared by a Sol-Gel Route", ©1998, with permission from Elsevier Science)

Composition 1)	K_{IC} 2) (MPa m$^{1/2}$)	P(%) 2)	σ_R 2) (MPa) 1600°C 2)	1450°C HP 3)	P(%) 3)	K_{IC} 3) (MPa m$^{1/2}$)
M	0.66	8	150	210	3.5	—
M^{30pl}	1.7	10	nm 4)	nm 4)	3	—
M_{10Zr}	1.04	2	280	280	2	—
M^{30pl}_{10Zr}	1.9	<2	280	340	2.5	—
A	—	2	300	340	<2	—
A^{30pl}	—	21	nm 4)	nm 4)	—	—
A^{30pl}_{10Al}	—	10	220	300	2	4.9

1)M : mullite (Baikalox 6944, Baikovsky, France), mean size 1.2µm ; A : alumina (AKP50, Sumitomo, Japan) mean size 0.6µm. pl : Single crystals of α-alumina (Atochem, France), hexagonal shape, mean size 5µm, thickness ≤1µm). Platelet content (30 wt%) and oxide content resulting from gel pyrolysis (10 wt%) are given in exponent and index, respectively. 2)Sintering temperature in air, 4 hours, 3)HP : hot pressing 16MPa, primary vacuum, 2 hours. 4)nm : Not measurable powders and particles are dispersed together in an anhydrous solvent before the addition of alkoxide. After mixing, pH controlled water is added under vigorous mechanical stirring which has to be maintained until a paste of thick consistency is obtained. The paste is then dried to give a flour-like powder. A sketch of the mixture cold pressing is given on Fig. 1 : during cold pressing, the viscous gel creeps to fill the interparticle voids. The faces of 1600°C air sintered pellets made under uniaxial compaction at 200MPa (4 minutes) are also shown in Fig. 1. The pellet made from a mixture of α-alumina (AKP50 powder, Sumitomo, Japan), and 30 wt% single crystals α-alumina platelets (Atochem, France), shows many cracks. On the other hand, a perfect pellet is obtained when the powder and platelets are embedded in a gel (≥10 wt%).

Comparison of the ultimate flexural strength, open porosity and toughness of various pellets is given in Table I. The densification hindrance induced by the platelet addition is straightforward. Gel embedding (zirconia, alumina or aluminosilicate precursor) significantly increases the toughness. Combining platelets ' addition and gel embedding facilitate the compaction (crack free pellets) and optimize ultimate strength as well as toughness[8].

MATRIX NET-SHAPE SINTERING AND MECHANICAL PROPERTIES OPTIMIZATION : THE INFILTRATION AND *IN SITU* GELATION OF SECOND PHASE PRECURSORS WITH CONTROLLED REACTIVITY

One of the main problems in the preparation of ceramic matrix composites is to achieve a low open porosity in the matrix, which implies matrix precursor penetration in the interfibre voids. In the case of weavable fibers, voids are a few microns or less in size, which makes infiltration of liquid precursors mandatory, in spite of their low ceramic yield. (gaseous precursors are rarely used for oxide compositions). The presence of a long fiber reinforcement, geometrically invariant, inhibits the coherent shrinkage of the matrix. In the case of particulate, 1D (long fiber) or 2D (textile) reinforcement, this phenomenon can be counterbalanced by hot-pressing : the in-plane shrinkage is counterbalanced by the thickness reduction. For woven fabrics of moderate thickness (≤3mm, typically), full impregnation by a mixture of alkoxides, which will be hydrolyzed and *in situ* gelated by reaction with atmospheric moisture, is possible. The fabric is embedded in a gel the viscosity of which assures, during the hot-pressing step, a homogeneous transfer of the applied pressure, preventing the fibers from breaking[9]. Furthermore, the choice of the gel composition allows control of the fiber surface degradation and of the thermal expansion mismatch[10]. By selecting the appropriate sol-gel precursors, the dwell temperature required to achieve good densification can be raised or lowered by about 100°C. This makes it possible to combine several kinds of impregnated/coated fabrics in the same composite in order to tailor some physical or chemical properties[11]. Examples are given on Fig. 2.

For continuous 3D fiber reinforced bodies, the idea of using liquid ceramic precursors for the impregnation of the fibrous preform originates in the preparation of C/C composites by pitch or phenolic resin infiltration. The slip-casting of a submicronic powder has been proposed, but this led to highly porous samples (open porosity close to 35-40%) exhibiting very poor mechanical properties and low protection of the fiber against corrosive atmospheres. Besides, the range between the consolidation temperature and the temperature at which the matrix shrinkage leads to a matrix cracking, within the geometric invariant preform or fabric, is small. Due to the fact that the achievement of a zero-porosity matrix is non-realistic, we will consider methods increasing the mechanical strength of a porous body. The first idea is to avoid cracks, to maintain a coherent matrix (maximization of the number of interparticulate bonds), the second one is to improve the strength of the interparticulate bonds and/or of the intergranular phase. The first requirement can be obtained by filling the voids between slip-cast

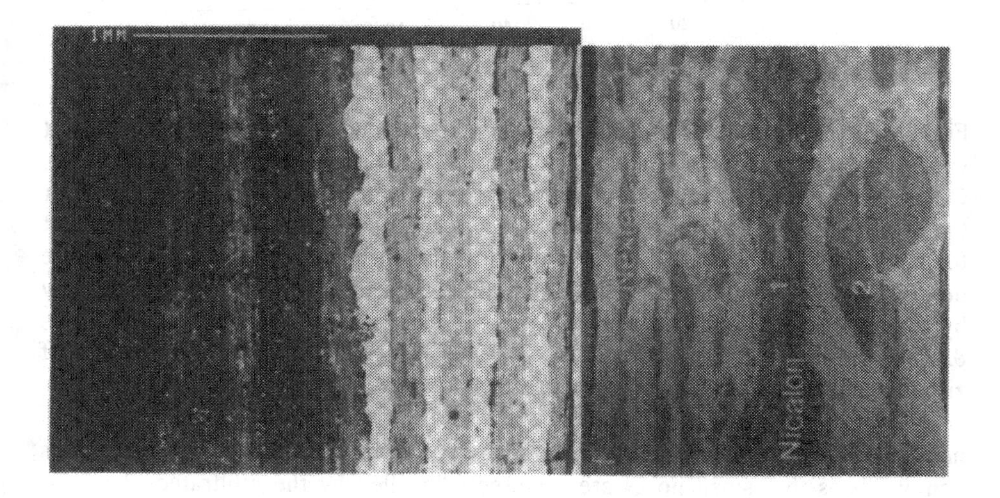

Fig. 2 : Examples of functionally graded composites combining *left*, mullite (grey) and zirconia (white) matrices, both reinforced with SiC Nicalon NLM202 fibers coated with an aluminosilicate interphase, and *right*, Nasicon matrix reinforced with 4 fabrics of Nextel 440 mullite and 2 fabrics of Nicalon NLM202 fibers. Bar : 1mm. (Courtesy of Materials Technology **10**(5/6) : 93-96[12]).

grains, with a ceramic precursor. Then, a thermal treatment transforms the precursor into a refractory phase. The resulting intergranular phase may form an inert or diffusion barrier between slip-cast grains and thus put the shrinkage off till higher temperatures[13].

Fig. 3 compares the evolution of the room-temperature flexural bending strengths of alumina monoliths prepared by slip-casting and sintering between 900 and 1400°C, and of the same samples after the following steps : **i)** 4 time infiltration by zirconium i-propoxide (or zirconium i-propoxide and aluminum s-butoxide, alternatively), **ii)** hydrolysis-polycondensation at 130°C in water and **iii)** thermal treatment at the same temperature as above (900-1400°C). The improvement of mechanical properties is due to the increase of the number of

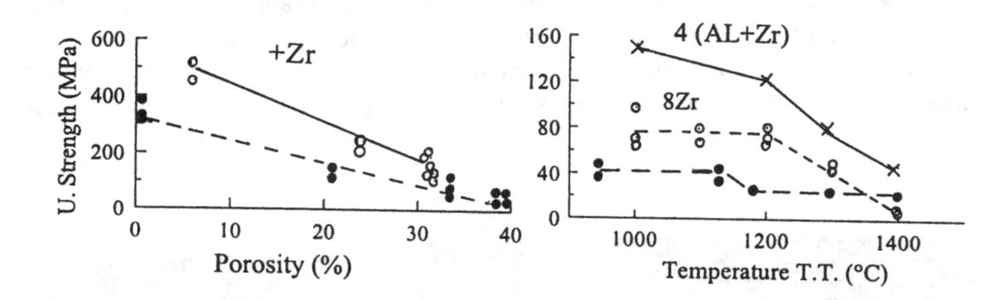

Fig. 3 : *Left*, room-temperature strength of slip-cast alumina monoliths sintered at various temperatures, plotted as a function of open porosity (dashed line). Comparison is made with data of monoliths post-infiltrated with a ZrO_2 precursor (solid line). *Right*, room-temperature strength versus thermal treatment temperature of a Novoltex Carbon preform slip-cast alumina matrix composite before (dashed line, black circles) and after cycles of post-infiltration, *in situ* hydrolysis-polycondensation and subsequent sintering at various temperatures: 8Zr, white circles, Zirconia precursor, 8 cycles; 4(Al+Zr), crosses, alternate use of zirconia and alumina precursors, 4 cycles.

necks between particles. However, comparison of the pore diameter distributions demonstrates that small pores are preferentially filled by the infiltrated phase, as expected from the capillarity and wetting behavior of alkoxides (Fig. 4). Coating of the internal wall of big pores is however evidenced by the left shift of all curves and confirmed by SEM investigations[8]. The infiltration process of a porous body using alkoxides is discussed in the next paragraph.

Composites made of an alumina matrix slip-cast within a 3D Novoltex Carbon (SEP, Le Haillan, France) preform, pore filled by **i)** alumina precursor (4-cycles of

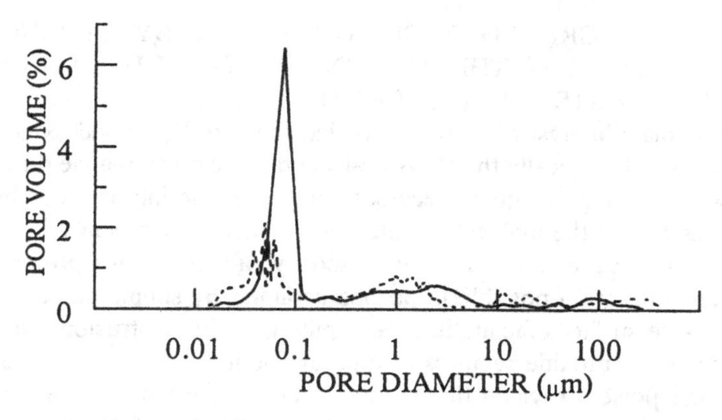

Fig. 4 : Pore distribution in slip-cast alumina monolith sintered at 1000°C before (solid line) and after post-infiltration with Al-butoxide, *in situ* hydrolysis-polycondensation and subsequent 1000°C sintering (dashed line)

aluminum-butoxide infiltration, which, due to a very high water sensitivity must be made in dry box, H_2O<20ppm and, then, subsequent hydrolysis) or **ii)** zirconia (4-cycles of zirconium-propoxide precursor-infiltration, in air) show similar mechanical properties (ultimate strength ~100MPa at R.T.) for a 1000°C thermal treatment, the porosity of the matrix being similar (35-36%)[13]. The sintering of slip-cast alumina submicronic powder takes place at temperatures ranging from 1000 to 1400°C by solid-state diffusion at the particle contacts. The shrinkage of the alumina xerogel issued from Al-butoxide hydrolysis takes place in the same temperature range. On the other hand, the full densification of zirconia xerogel requires temperatures higher than 1400°C. Consequently, the sintering of the alumina matrix post-infiltrated with zirconia precursor is delayed up to 1300°C[13].

As reported above, the methods to optimize the mechanical properties of a 3D fiber reinforced porous matrix composite are the prevention of cracking and the achievement of strong interparticle bonds. The latter has been achieved by post-infiltration with a second precursor, using **i)** aluminum-silicon ester ($(Bu)_3Al-O-Si(O(Et)_3)$, which, contrarily to the first zirconia-precursor, can react with the matrix to give an intergranular cement or, **ii)** aluminum s-butoxide and zirconium i-propoxide, alternatively or mixed together, to obtain a nanocomposite intergranular phase. The room-temperature flexural strength was multiplied by five times after 5-cycles of post-infiltration, *in situ* hydrolysis-polycondensation and 1000°C heating. Zirconium i-propoxide was used for the first 4-cycles and aluminum-silicon ester for the last one. Furthermore, measurements of the flexural strength at 1200°C and at 1300°C show that the mechanical properties are increased by 6.

FUNCTIONALLY GRADED MICROSTRUCTURE BY FILLING OF A GRADED POROSITY : THE USE OF A LOW VISCOSITY LIQUID PRECURSOR AND ITS *IN SITU* GELATION

One of the main interest of alkoxides is that they are liquid and exhibit a rather good ceramic yield : typically the alkoxide-to-ceramic conversion yield ranges from 20 to 30 wt%. Mixing of liquid precursors promotes the intimate combination of various precursors at the molecular scale[1-4] and allows adjustment of the viscosity to voids ' filling requirements. The low viscosity of zirconium i-propoxide (~0.5 poise) makes infiltration possible in porous ceramics by simple dipping, the good wetting of oxide surfaces facilitating the capillarity driven intrusion. On the other hand, aluminum s-butoxide requires heating at about 80°C to lower the viscosity from 13 to 0.5 poise. Provided that the infiltration is not hindered by water traces in the pores, which can induce their warping, the number of cycles needed to fill the porosity can be calculated from the ceramic conversion yield. The infiltration law is :

$$P_{n+1} = P_n\left(1-\frac{rd_a}{d_p}\right) = P_0\left(1-\frac{rd_a}{d_p}\right)^{n+1}$$

where P_o : initial porosity of the body, P_n = porosity after n cycles of infiltration-hydrolysis-polycondensation and thermal treatment, d_a : alkoxide density, d_p : density of the oxide resulting from gel pyrolysis and r the alkoxide-to-ceramic conversion yield (in weight). For instance, using zirconium propoxide and aluminum s-butoxide, we obtained $P^{Zr}_{n+1} = P_0(0.946)^{n+1}$ and $P^{Al}_{n+1} = P_0(0.936)^{n+1}$. The number of Zr-propoxide infiltration cycles needed to lower the porosity by 5% would be 6, as verified experimentally. On the other hand, the experimental law measured after Al s-butoxide infiltration deviates from the theoretical prediction due to the difficulty to avoid water traces within a porous body. Pressure assisted infiltration improves the efficiency[13].

Comparison of the ultimate strength increase as a function of the porosity filling shows that a plateau is obtained after 4 cycles and that no profit is obtain up to 10 cycles. On the other hand, large differences are observed depending on the used alkoxide or alkoxide combination. This behavior can be related to the preferential filling of the small pores as demonstrated by measurement of the pore distribution (Fig. 4).

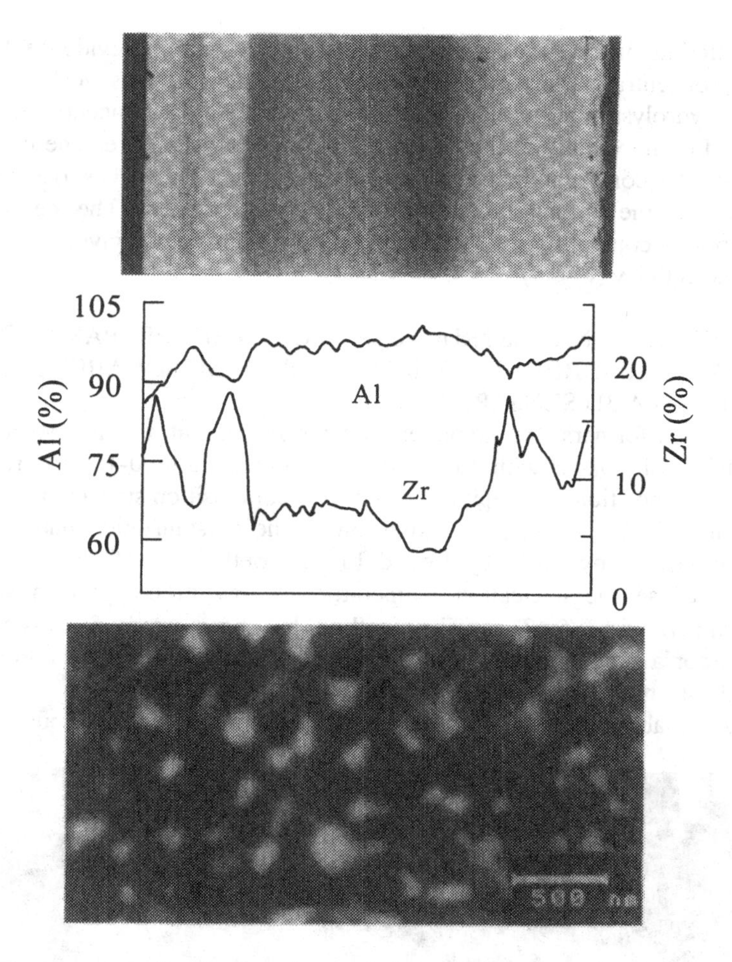

Fig. 5 : Back-scattered scanning electron micrographs of an alumina monolith after 4 cycles of zirconium i-propoxide infiltration, hydrolysis-polycondensation and firing at 1400°C. Plot of the Al and Zr concentrations (wt%) across the sample is given (scanned length : 2mm). Zirconia appears in white between slip-cast alumina grains.

Electron micrographs of polished sections of the slip-cast alumina monoliths sintered at a temperature comprised between the onset (1000°C) and the end (1400°C) of the sintering show a non-uniform distribution of the porosity (Fig.5). The larger pore concentration is observed in the upper region of the monolith, in contact with the slurry, and the lower in the vicinity of the lower surface, in contact with the filtering paper. Back-scattered electron micrographs evidence that the zirconium concentration across the monolith thickness after several cycles of infiltration, hydrolysis and pyrolysis is heterogeneous. Higher concentration waves are observed in the vicinity of the border of the low porosity core. The number of waves seems to correspond to the number of cycles. This gives rise to wave concentration of the zirconia functionally graded reinforcement. The measurement of the zirconium content in low and high concentration regions gives about 7 and 20 wt%, respectively (Fig. 5).

TAILORING THE NANOSTRUCTURE FOR THE PREPARATION OF MICROWAVE ABSORBENT : THE USE OF A MESO/ MICROPOROUS (XERO)GEL AS A 3D SUBSTRATE

In a gel, gel-formers are homogeneously dispersed at the molecular scale. Heterogeneity will appear with nucleation, typically in the 600-1000°C range for usual oxide compositions. Precipitates can have a composition similar to that of the matrix or not. In both cases, the extension of the crystallization and the grain growth are limited by the very low diffusion coefficients of the constitutive elements, because the nucleation temperature is low in comparison with the melting temperature (≤0.5Tm). This method has been used, for instance, to disperse zirconia nanoprecipitates in a mullite matrix, in order to improve the mechanical strength of SiC fiber reinforced composite matrix[9] : zirconia precipitates at about 1000°C, while pure zirconia nucleates at about 500°C[14].

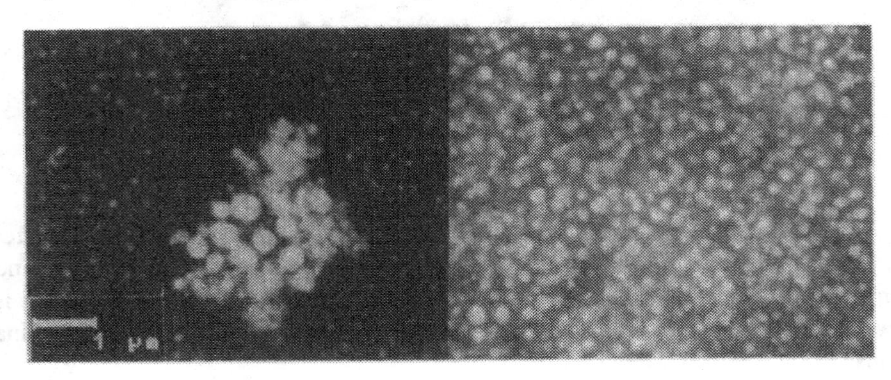

Fig. 6 : Backscattered electron scanning micrographs of cobalt-$Al_2O_3.2SiO_2.0.7B_2O_3$ composites prepared by mixing (a) and infiltration (b) routes. Materials have been thermally treated at 1100°C under H_2. (bar : 1μm) (Reprinted from Ph. Colomban *et al.* in Nanophase and Nanocomposite Materials II, edited by S. Komarneni, J.C. Parker and H.J. Wollenberg, pp. 451-462, Mater. Res. Soc. Proc. 457, 1997[15]).

However, this method is not versatile. Another route has been explored. Gels and xerogels exhibit high meso-, and sometimes micro-porosity. Thus, the polymeric porous backbone can be used as a substrate for a second precursor. The densification of the host framework will entrap the second precursor, adsorbed on the pore walls or logged within the pores, which will be transformed in the desired phase on heating. Comparison is given between the so-called mixing route, where the alkoxide mixture leading to the gel host framework is hydrolyzed using a transition metal nitrate solution, and the infiltration route where the gel is first formed, more or less dried (between 40 and 600°C) and then dipped in the metal precursor aqueous solution. Comparison of the electron micrographs have shown that the best dispersion is achieved by the infiltration route[15] (Fig. 6). Using microporous (xero)gels as host matrices the metal particles size ranges from 5 to 20nm for a 600°C thermal treatment under H_2, and can reach up to 100nm using a mesoporous host matrix. Combined infiltrations with different metal and alloy precursors and subsequent H_2 thermal treatments can tailor the microwave absorption[16].

CONCLUSION

The examples reported in this review show that the sol-gel routes offer many tools for the control of the microstructure and, in some cases, the nanostructure of multiphase materials, provided that the gel properties are well understood and that the thermal treatment temperature remains sufficiently low to avoid abnormal grain growth.

ACKNOWLEDGEMENTS

The author is thankful to Dr. E. Mouchon, Dr. E. Bruneton, D. V. Vendange, Dr. M. Wey for their contributions and to Dr. M. Parlier for helpful discussions.

REFERENCES

[1]L.C. Klein (ed.), "Sol-Gel Technology", Noyes Publications, Park Ridge, 1988.

[2]Ph. Colomban, "Gel Technology in Ceramics, Glass-Ceramics and Ceramic-Ceramic Composites", *Ceramics Int.* **15** 23-50 (1989).

[3]C.J. Brinker and G. Scherer, "Sol-Gel Sciences : The Physics and Chemistry of Sol-Gel Processing", Academic Press, New York, 1990.

[4]S. Sakka and K. Kamiya, "Glasses from Metal Alcoholates", *Journal of Non-Crystalline Solids* **42** 403-20 (1980).

[5]M. Kulig, W. Oroschin and P. Greil, "Sol-Gel Coating of Silicon Nitride with Mg-Al Oxide Sintering Aid", *Journal of European Ceramic Society* **5**[4] 209-17 (1989).

[6]J. Wang, R. Paramoon, C.B. Ponton and P.M. Marquis, "An Application of Sol-Gelation in the Dispersion Mixing of Ceramic-Matrix Composites", *Journal of Material Science Letters* **11** 807-09 (1992).

[7]Y. Hirata, S. Matsushita and Y. Ishihara, "Colloidal Processing and Mechanical Properties Whisker Reinforced Mullite Matrix Composites", *Journal of American Ceramic Society* **74**[10] 2438-42 (1991).

[8]J.L. Lagrange and Ph. Colomban, "Double Particle Reinforcement of Ceramic-Matrix Composites Prepared by a Sol-Gel Route", *Composite Science and Technology* **58**[4] (1998) in press.

[9]Ph. Colomban, E. Bruneton, J.L. Lagrange and E. Mouchon, "Sol-Gel Mullite Matrix-SiC and Mullite Woven Fabric Composites with or without Zirconia Containing Interphase : Elaboration and Properties", *Journal of European Ceramic Society* **16**[2] 301-14 (1996).

[10]S. Karlin and Ph. Colomban, "Raman Study of the Chemical and Thermal Degradation of as-Received and Sol-Gel Embedded Nicalon and Hi-Nicalon SiC Fibres Used in CMC's", *Journal of Raman Spectroscopy* **28** 219-28 (1992).

[11]Ph. Colomban, "Sol-Gel Control of the Micro/Nanostructure of Functional Ceramic-Ceramic and Metal-Ceramic Composites", *Journal of Material Research* **13**[4] 803-11 (1998).

[12]Ph. Colomban, "Process for Fabricating a Ceramic Matrix Composite Incorporating Woven Fibers and Materials with Different Compositions and Properties in the Same Composite", *Technological Advances, Material Technology* **10**[5/6] 89-103 (1995).

[13]Ph. Colomban and M. Wey, "Sol-Gel Control of Matrix Net-Shape Sintering in 3D Fibre Reinforced Ceramic Matrix Composites", *Journal of European Ceramic Society* **17**[12] 1475-83 (1997).

[14]Ph. Colomban and L. Mazerolles, "Nanocomposites in Mullite-ZrO_2 and Mullite-TiO_2 Systems Synthesized through Alkoxide Hydrolysis Gel Routes", *Journal of Material Science* **26** 3503-10 (1991).

[15]Ph. Colomban and V. Vendange, "Sol-Gel Routes Towards Magnetic Nano Composites with Tailored Microwave Absorption", pp. 451-62, in *Nanophase and Nanocomposite Materials II*, edited by S. Komarneni, J.C. Parker and H.J. Wollenberger, vol. 457, Materials Res. Soc., Warrendale , 1997.

[16]V. Vendange, E. Tronc and Ph. Colomban, "Towards microwave absorbing Co(Fe)-oxide nanocomposites", *J. Sol-Gel Science and Technology* **11**[3] (1998) in press.

MICROSTRUCTURAL EVOLUTION ON SINTERING IN Y_2O_3-DOPED ZrO_2 SYSTEM

N. de la Rosa-Fox, M. Piñero, A. Santos[b], C. Jiménez-Solís, C. Barrera-Solano and L. Esquivias.
Dpto. Física Materia Condensada, Universidad de Cádiz
[b]Dpto. Cristalografía, Universidad de Cádiz
Aptdo. 40, 11510 Puerto Real (Cádiz) SPAIN;
E. Hoinkis.
Berlin Neutron Scattering Center. Hahn-Meitner Institut
D-14109 Berlin. GERMANY

ABSTRACT

ZrO_2-Y_2O_3 ceramic powder was obtained by controlled hydrolysis of alkoxides in water-free atmosphere at room temperature. Typical compacts contain local variation in chemical composition, grain size, pore volume, residual stress, etc. We used Small Angle Neutron Scattering (SANS) to study the microstructure evolution in powder agglomeration in compacts of ceramics during sintering process on the 1-100 nm length scale. Powder scattering curves for particles of 50 nm mean size show features of a two-aggregation level. On the other hand, compacts of ceramics reveal the scattering from the pore/grain interface with a coarse porosity. The small closed pores of 5-10 nm size produce an important interference effect on the neutron scattering, which disappears when a 90 % relative density is reached due to their collapse.

1. INTRODUCTION

ZrO_2-Y_2O_3 ceramic system is widely studied because of its noteworthy mechanical performance and low thermal conduction. These properties enable to

To the extent authorized under the laws of the United States of America, all copyright interests in this publication are the property of The American Ceramic Society. Any duplication, reproduction, or republication of this publication or any part thereof, without the express written consent of The American Ceramic Society or fee paid to the Copyright Clearance Center, is prohibited.

use them as refractory materials to work under stress, as those produced by thermal expansion.

Special attention has been focused on the preparation of YSZ (Yttria Stabilised Zirconia) ceramic powders with a predesigned size distribution in the nanometric scale. Fine-grained sintered ceramics can now be produced using commercially available ultrafine powders, with particle sizes less than 100 nm or even 10 nm[1,2], which can be sintered to full density at relatively low temperatures. Though several wet chemical methods have been developed to obtain such sinterable powders, a critical parameter that affects final density is the state of agglomeration. This, can be lessened or even eliminated by alcohol washing treatment, thus enhancing densification and microstructure development during sintering of the ceramic body. However, critical microstructural defects may originate along the powder and compact forming processes, thus limiting the reliability of the materials. Typical compacts contain local variations in chemical composition, particle size, porosity, residual stress, etc. In this way the SANS study may be focused to improve the production of fully-dense YSZ ceramics.

2. SAMPLE PREPARATION FOR SANS EXPERIMENTS

Highly dispersed Zirconia powders containing 3 and 6 mol % of yttria were prepared by controlled hydrolysis of Zr-isopropoxide and Y-acetate suitable mixtures in highly diluted ethanolic solutions. All reactions took place in N_2 atmosphere at room temperature. After drying at 100 °C the amorphous powders were calcined at 370 °C. Subsequent thermal treatment at 600 °C induced the formation of the tetragonal stabilised crystalline phase. Green compacts were then heat treated at a rate of 5 °C/min in the 1000-1600 °C temperature range with 200 °C during 4 hours for each.

Powder compacts were prepared by uniaxial pressing under a pressure of 110 MPa in 10 mm diameter and 1 mm thick pellets. SANS measurements were carried out in the V4 experimental station at the BENSC facility of the Hahn-Meitner Institut (Berlin). The neutron beam wavelength was selected at 6.02 Å, and the scattering data were recorded by using sample-detector distances of 16, 4 and 1 m to cover a scattering vector range (q) of 0.036-3.6 nm^{-1}. No evidence of multiple scattering was detected since the neutron transmission was higher than 70 %.

Standard programs were used for the data reduction, e.g. normalization of the efficiency of the two-dimensional detector, masking of the detector cells near the beam stop, and substraction of the scattering by the empty cell and of the electronic background. The macroscopic coherent scattering cross section

$d\Sigma(q)/d\Omega$ in units of inverse centimeters was calculated by using the scattering data from H_2O. For convenience $d\Sigma(q)/d\Omega$ is represented here by I(q)

Crystalline phases were identified by X-Ray Diffraction (XRD) and Differential Thermal Analysis (DTA) techniques.

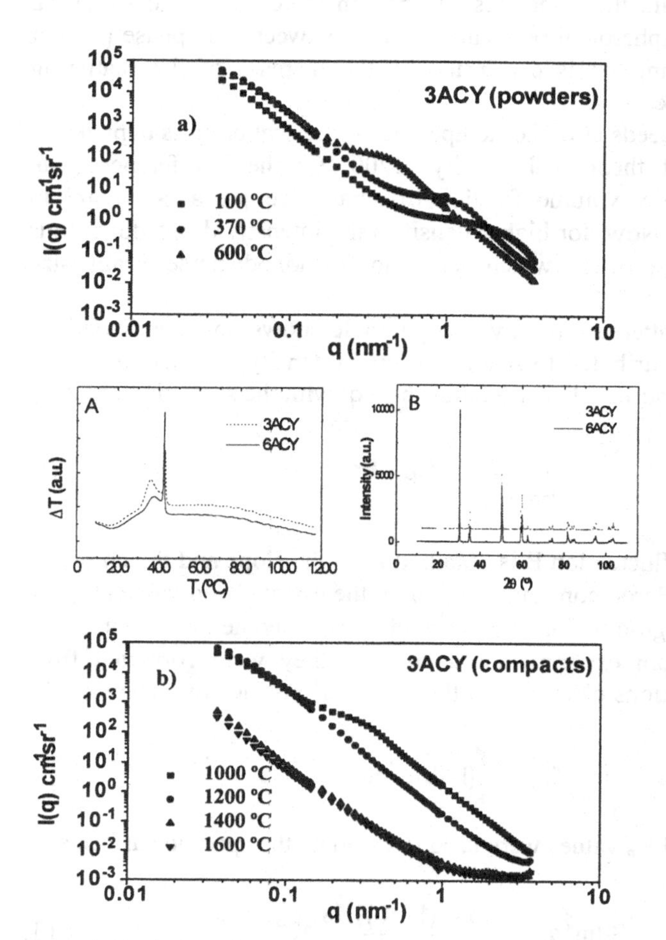

Figure 1. Evolution of the corrected scattered intensity with heat-treatment a) 3ACY powders, and b) 3ACY compacts. DTA curves of 3ACY and 6ACY samples at 100°C are shown in the inset A, and the diffractograms corresponding to 3ACY and 6ACY at 1400 °C are presented in the inset B.

3. RESULTS

DTA curves obtained from 3 and 6 mol % yttria-doped zirconia powders (3ACY and 6ACY respectively) show two exothermic peaks at temperatures of 380 °C and 425 °C (inset A in Fig.1). The first one is attributed to decomposition of organics, and the sharp peak indicates crystallisation of zirconia in both tetragonal and cubic crystalline phases as revealed by XRD techniques (inset B in Fig 1). Fig.1a. shows the coherent scattering cross section in a standard log-log plot for the 3ACY powders, the behavior of the corresponding curves for 6ACY powders being almost the same. The different heat treatments for the 3ACY are pointed out at the outlined temperatures. The curves evolve from the high to the low-q region as a knee. As the temperature is increased from 100 °C to 600 °C, the interference contribution from the pore/grain interface can be discerned at intermediate q as a peak, which is shifted to lower q values for higher temperatures. At this point these features can be considered as a two structural level, where interacting spheroidal pores are situated between nanophase powder agglomerates. This assumption is confirmed by the increase of the scattering intensity with temperature.

As densification proceeds at higher temperatures, the scattering is depicted in Fig.1b. For a 65 % of theoretical density (1200 °C) the interference peak dissapears due to the low volume fraction of small size pores, as found in previous SANS studies[3]. Now, for higher densities the intensity drops down thus revealing the sintering process which leads to a monodisperse grain size distribution.

The decay of the scattered intensity at high angle shows positive deviations from the Porod's law[4,5] attributed to scattering length density fluctuations which leads to not constant but surface linear relationship Iq^4 with the scattering angle:

$$\lim_{q \to \infty} Iq^4 = K_P + Bq^4 \tag{1}$$

where the magnitude of fluctuation B is obtained from the slope and the corrected asymptotic value of the Porod constant is given by the extrapolated value at $q = 0$ from a linear fit in the high-q region. This correction permits the evaluation of the integrated intensities from experimental data, once they were corrected from scattering length fluctuations observed on the Iq^4 vs. q^4 profiles by means of the relationship[4,6]:

$$Q_0 = \int_0^{\infty} (I - B) q^2 dq \tag{2}$$

The resulting Q_0 and K_P values were used to evaluate the specific surface area from relationship:

$$S'(m^2 g^{-1}) = \frac{\pi \Phi (1 - \Phi)}{\rho (g \cdot cm^{-3})} \frac{K_P}{Q_0} 10^3 \tag{3}$$

Where ρ is the density geometrically measured, and Φ is the relative density estimated from theoretical (6.03 $g \cdot cm^{-3}$). As expected, the pore surface area decreased during densification. Its temperature dependence is presented for the 3ACY samples in Fig.2.

The Guinier approximation of non-interacting particles gives the radius of gyration Rg for the scatterers, from the relationship:

$$I(q) = I(0)\exp(-q^2R_g^2/3) \tag{4}$$

This approximation holds only for samples heat treated at ≤1000°C, in the region of small q values where qRg <1, as it can be seen in Fig.1a.

For higher q-regions the Debye function, characterized by the correlation length, ξ, and Porod´s behavior for high-q values, which gives:

$$I(q) = \frac{I(0)}{(1+\xi^2q^2)^2} \tag{5}$$

describes a polydisperse porous solid, being useful for all the scattering curves in the experiment (Fig. 1a and 1b). From those analysis two scatterer sizes have been found as surface area decreases in samples treated under 1000°C. This confirms the assumptions made before on the existence of two-level structure at early stages of sintering.

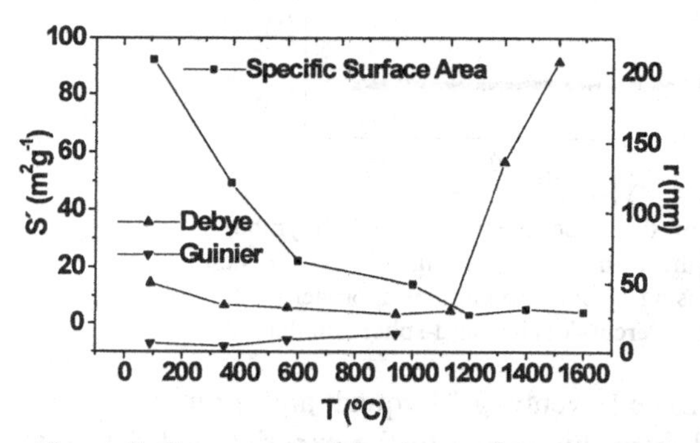

Figure 2. Temperature dependence of Porod specific surface area and particle size for the scatterers evaluated from Guinier and Debye approaches for ZrO_2-3 mol % Y_2O_3 samples.

Debye approximation applied at 1200°C, 1400°C and 1600°C sintering steps, reveals the grain growth to form a 200 nm grain size with some remaining closed porosity of 4 m^2g^{-1} (Fig 2). This is accompanied by an increase in the compactness from 65% to 90% theoretical density.

The total intensity is a convolution of the form factor P(q) and structure factor S(q) as:

$$I(q) = nP(q)\cdot S(q) \tag{6}$$

where n is the number density of the scatterers. The scattering intensities were then calculated by using the Guinier and Debye approaches to obtain P(q). The structure factor, S(q), was calculated by the Percus-Yevick hard spheres model, as can be seen in Fig. 3, using the expression of Ashcroft and Lekner[7] which relates the value of S(q) to both the hard sphere diameter, HD, and the packing fraction,

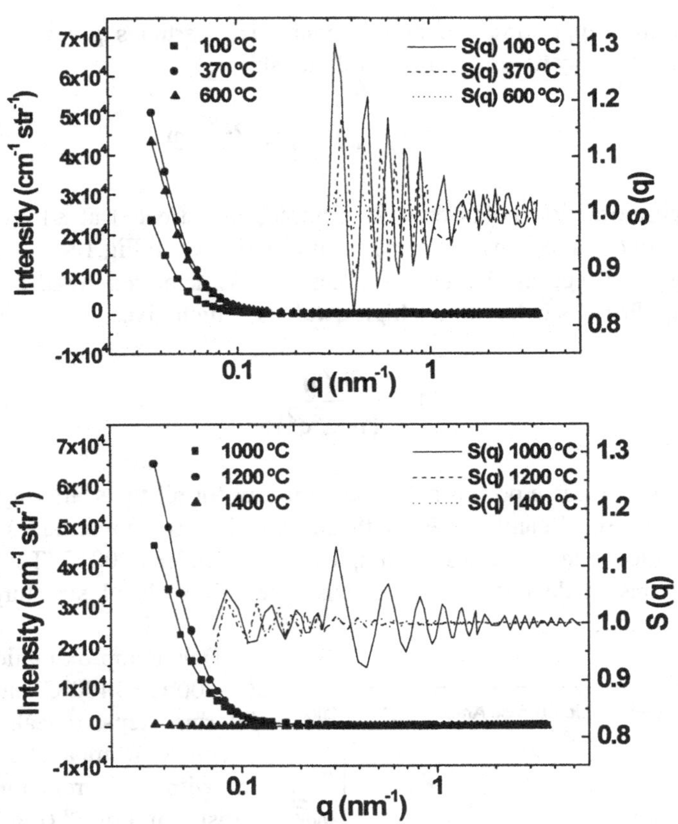

Figure 3. Deconvolution of the experimental intensity (symbols) from 3ACY samples into the form factors (straight lines), by non linear least squares fitting using Debye and Guinier approaches, and the structure factors using the Percus-Yevick hard-sphere model.

η. Non-linear least squares based on the Levenberg-Marquadt algorithm was used to fit the SANS patterns generated from this model to the experimental data by minimizing the Eq. (6). The results are collected in Table 1.

Fig. 3 shows how S(q) deviations from one decrease rapidly at early stages of sintering, indicating a low interparticle interference term caused by a reduction of powder agglomerates as the sintering progresses. This is corroborated by the decrease in packing fraction and HD values obtained from Percus-Yevick in the 100ºC-1000ºC temperature range (Table 1).

Table I. Hard sphere diameter HD and packing fraction η of elementary particles according to Percus-Yevick model for 3ACY samples at the outlined temperatures.

Temperature (°C)	HD (nm)	η
100	528	0.46
370	421	0.41
600	297	0.26
1000	297	0.20

4. CONCLUSIONS

SANS experiments resolve similar powder aggregation and compact microstructure for both 3 and 6 mol % yttria-stabilised zirconia studied samples. This technique permits to know some parameters of the inner part of the grain boundary.

Ceramic powder is built by a two level structure. The first level is formed by clusters of about 50 nm size. Between them a fine porosity of 5 to 10 nm size exists as the second level. A decrease of the specific surface area indicates the collapse of such porosity.

On the contrary, compact ceramics are only formed by one level structure. The sintering process reveals the grain growth to form a 200 nm grain size with some remaining porosity (closed) of 4 m^2g^{-1}. This process is accompanied with an increase in the compactness from 65 to 90 % theoretical density (in order to be seen by SANS probe).

By using classical models (Debye, Guinier, Percus-Yevick) experimental SANS intensities were fitted by a non-linear least squares procedure. The results of the form factor P(q) confirm the polydisperse distribution of the ceramic grains and the homogeneous coarse porosity between. On the other hand, the structure factor S(q) accounts for the aggregation process in the second structural level, revealing a low powder agglomeration.

Neutron scattering measurements in Berlin has been supported by the TMR/LSF access program (ERBFMGECT950060) of the European Commision.

REFERENCES

[1] K. Haberko, "Characteristics and Sintering Behaviour of Zirconia Ultrafine Powders", *Ceramurgia Int.*, **5** [4] 148-154 (1979)

[2] B. Fegley, P. White, H.K. Bowen, "Processing and Characterization of ZrO_2 and Y-doped ZrO_2 Powders", *Am. Ceram. Soc. Bull.*, **64** [8] 115-1120, (1985).

[3] S. Krueger, A.J. Allen, G.Skandan, G.G. Long, H. Hahn, H.M. Kerch, "Small-Angle Scattering Methods for Studying the Sintering of Nanophase Ceramic Compacts", *Mat. Res. Soc. Symp. Proc.* Vol 376, 359- 364 (1995)

[4] W. Ruland, "Small-Angle Scattering of Two-Phase Systems:Determination and Significance of Systems Deviations from Porod's Law" , *J. Appl. Cryst.*, **4** , 70-73 (1971)

[5] J. T. Koberstein, B. Morra, R.S. Stein, "The Determination of Diffuse-Boundary Thicknesses of Polymers by Small-Angle Scattering" *J. Appl. Cryst.*, **13**, 34- 45 (1980).

[6] A. F. Craievich, "SAXS Study of the Porous Fractal Structure of Tricalcium Silicate Dry Gels", *J. Appl. Cryst.*, **20**, 327-329 (1987)

[7] Y. Waseda, pp.22-23 in "The Sructure of Non-Crystalline Materials", McGraw-Hill, New York, 1980.

KEYWORD AND AUTHOR INDEX

9 781574 980639